CRITICAL ACCLAIM FOR
NAKED EARTH

"There's a whole new world beneath our feet, as turbulent and varied as anything on the surface. In *Naked Earth*, Shawna Vogel takes us on a fast-paced tour of this hidden realm, marrying luminous science writing with vivid accounts of the scientists who are trying to plumb it." —*Science* magazine

"A lively tour through the planet's dynamic inner landscape."
—*Boston Herald*

"Brings a weighty subject vibrantly to life. . . . This is top-notch science journalism." —*Publishers Weekly* (starred review)

"Shawna Vogel's precise and visual prose conjures up startling pictures of a vast underworld on an exhilarating journey to the center of the earth and back. Along the way, the reader effortlessly grasps the challenges scientists face, and the many tricks they employ, to explore the mysterious world beneath our feet." —Lauren S. Aguire, producer, *Nova*

"Fast-paced and breezy . . . Vogel succeeds marvelously in avoiding the specialists' jargon and providing a clear description that should satisfy the intelligent nonscientific reader."
—*New York Times Book Review*

"Superbly written and informative . . . dynamic and unpretentious . . . Vogel's infectious enthusiasm and lucid prose make this debut an engaging page-turner." —*Kirkus Reviews*

SHAWNA VOGEL holds a B.S. from MIT, where she was awarded the prestigious Knight Science Journalism Fellowship. A former editor at *Discover* magazine, she writes for various popular science magazines, including *Scientific American*, *Earth*, and *Discover*. Ms. Vogel currently lives in Boston.

NAKED EARTH

THE NEW GEOPHYSICS

SHAWNA VOGEL

A PLUME BOOK

PLUME
Published by the Penguin Group
Penguin Books USA Inc., 375 Hudson Street, New York, New York 10014, U.S.A.
Penguin Books Ltd, 27 Wrights Lane, London W8 5TZ, England
Penguin Books Australia Ltd, Ringwood, Victoria, Australia
Penguin Books Canada Ltd, 10 Alcorn Avenue, Toronto, Ontario, Canada M4V 3B2
Penguin Books (N.Z.) Ltd, 182–190 Wairau Road, Auckland 10, New Zealand

Penguin Books Ltd, Registered Offices: Harmondsworth, Middlesex, England

Published by Plume, an imprint of Dutton Signet,
a division of Penguin Books USA Inc.
Previously published in a Dutton edition.

First Plume Printing, May, 1996
10 9 8 7 6 5 4 3 2 1

The Library of Congress has catalogued the Dutton edition as follows:

Vogel, Shawna.
Naked Earth : the new geophysics / Shawna Vogel.
p. cm.
ISBN 0-525-93771-4 (hc.)
ISBN 0-452-27162-2 (pb.)
1. Earth—Core. 2. Earth—Internal structure. I. Title.
QE509.V63 1995
551.1—dc20
94-23784
CIP

Printed in the United States of America
Original hardcover design by Eve L. Kirch

CONTENTS

INTRODUCTION: VISIONS OF HELL

High on the northern coast of Europe, on an elbow of land called the Kola Peninsula, looms the Eiffel Tower of the Russian tundra, a boxy brick derrick soaring thirty stories above the horizon. For over two decades geologists have labored under this monolith, withstanding the bitter cold and shifting daylight hours of life above the Arctic Circle. They are gouging a hole into the Russian bedrock, drilling into regions of the planet that have never before been seen.

The Kola Hole, now stretching more than seven and a half miles into the ground, is the deepest hole on earth. It surpasses even the natural abysses of the ocean floor. Through layer upon layer of rock, the hole penetrates ever farther into the earth's history. Passing through rock that formed as the first multicelled organisms were roaming the earth, it now reaches a time when the atmosphere as we know it had not yet evolved. All told, 1.4 billion years of history line the walls of the Kola Hole, a record in stone that has served up a wealth of surprises to Russian geologists—but none so strange as the incident that was reported in an American tabloid not long ago.

It began, according to one member of the drill crew, when they noticed their drill bit was heating up rapidly. As the crew pulled it up to see what the matter was, out of the hole flew a hideous, gargoyle-like beast in a cloud of bituminous smoke. This inky creature squawked and shrieked and fluttered out of sight. Then down the hole the crew lowered a sensitive microphone. "We could hardly believe our ears," one scientist reportedly said. "We heard a human voice, screaming in pain. Even though one voice was discernible, we could hear thousands, perhaps millions in the background." Could it be that they were hearing the cries of legions of wicked souls? Had the Kola Hole pierced the house of Satan himself?

Readers were left to draw their own conclusions. But the tabloid, as evidence of its story's veracity, pointed out that under Communism, religion and all of its gods and demons had been suppressed for seventy years. So if a Russian crewman reported the incident, then it must have been true, because he couldn't possibly have made up such a tale!

When I first heard this story, I was standing with a group of geophysicists after a long day of conference talks. Drinks in hand, the crowd around me erupted with groans of exasperation, the kind usually reserved for terrible puns. Not only was this story a fantastic piece of fiction to them, but a bittersweet reminder that tall tales about underworld civilizations still haunt their profession. Along with flat-earth theories and little green men on Mars, such wild speculations about the inner earth would die a natural death if it were up to the geophysicists in this room, for these researchers have spent their careers straining to hear the noises that come from the earth's deeps.

Through these efforts they are now beginning to see what

is really inside the earth—not creatures and kingdoms, but a fantastic subterranean landscape that rivals any in our imaginations. Bit by bit as geophysicists peer through the depths of earth, their revelations about how the interior looks are giving way to an even more valuable insight: how the planet works.

Researchers are discovering how the planet's layers, from the crust and mantle to the core, all work together in this engine we call the earth. They have learned not only that all the continents on earth once formed a giant landmass known as Pangaea, but that the presence of this supercontinent on the surface 180 million years ago is now stored like a memory in the rock of the inner earth and may even be dictating where enormous floods of magma erupt on the surface today. Geophysicists have also uncovered a connection between such magma floods and one of the earth's great mysteries: why the magnetic poles have, many times in the past, reversed their positions, north for south.

Yet no matter how scintillating their science or how deep into the earth it reaches, geophysicists exploring the earth's interior must still contend with the far-fetched notions of the past. Giant moles, three-headed dogs, old friends, kings and queens and imprisoned, malicious or merely sleeping gods have all dwelt in the belly of the earth in popular mythology. The folklore of the Kukis, a community living on the outskirts of the Himalayas, holds that the periodic quaking of the earth is caused by a race of beings that inhabit its interior. The Kukis' legends tell that these subterranean beings stomp their feet to see if people are still living on the earth's surface. So whenever the Kukis feel the ground start to shake, they shout, "Alive! Alive!" until the trembling ceases.

In at least one tale, the earth itself is the living creature. That is the premise of "When the World Screamed," a short story published in 1929 by Sir Arthur Conan Doyle, the British author best known as the creator of Sherlock Holmes. After Holmes, perhaps Doyle's most enduring character is Professor George Challenger, a brusque, bullheaded adventurer and scientist who, in this particular story, digs a hole eight miles into the crust. At the bottom of the hole is a slimy, pulsating mass that Challenger believes to be the tender tissue of the living earth. Too vain to be merely a flea on this animal's back, the professor's aim is to give it a poke, "to let the earth know that at least one person, George Edward Challenger, calls for attention."

The title of the story gives away the ending. Before a crowd of spectators, a sharp iron bore is sent down the shaft into the skin of the beast. The earth is indeed alive—and furious at being stabbed. It shrieks. It sends a gusher of foul, tar-like dreck up the hole, soaking the spectators. Then it obliterates all of Professor Challenger's digging by sealing its own wound in a rumble of collapsing rock. The planet had spoken. No one, not even Challenger, would see what lay beneath its stony carapace.

Yet decades earlier, readers had already gotten a vivid fictionalized look at the inner earth from another science-fiction author. Jules Verne had such a mastery of the scientific thinking of his day that his early fans scarcely knew where fact left off and fantasy began. Millions of readers traveled along with the brilliant Professor Lidenbrock, his nephew and a guide through the mouth of an Icelandic volcano on a fascinating *Journey to the Center of the Earth.*

With only their headlamps to light the way, these adventurers trekked past layer upon breathtaking layer until finally the tunnel they were following opened onto an enormous cavern. The entire void glistened with the strange lights of underground electric currents. These lights danced across the surface of a murky body of water which the travelers named the

Central Sea. In it swam species of fish that no longer dwelt in the surface world, and along its shores lay the chalky skeletons of prehistoric mastodons. Even a human skull turned up in the fossil heap, convincing the travelers that they were not the first to pass this way.

Indeed, many mythical explorers have made this same journey. Verne's subterranean tale stems from a centuries-old tradition of stories about the inner earth. These tales satisfy a natural human need to understand the world on which we live, and thus they have become an integral part of the many varieties of religion and culture that we have subscribed to over the years.

Think of the Mediterranean basin, cradle to Western civilization and, then as now, the frequent site of both earthquakes and volcanoes. In ancient times storytellers and traders spread the word of these events from one community to the next, and elders passed the news down to their children. Gradually these cataclysms seeped into the religious doctrine of the region's inhabitants. What could be better proof of God's presence, after all, than the wholesale trembling of the earth? What could be better evidence of His wrath than the eruptive blast of a volcano such as Mount Etna?

Since that early intermingling of the earth's power with religion, the boundary between science and myth has been blurred. For centuries, hell was conceived of as a fiery void in the earth's interior, where the damned suffered everlasting torment. As the years progressed, poets like Virgil and Dante built upon this belief. In the *Aeneid* and *The Divine Comedy* they invented vivid underworld landscapes of torturing demons, burning deserts and bottomless pits. Perhaps these images endured through the years because they struck a chord in many people's hearts. Volcanoes and earthquakes are such violent events that it may have seemed they could only be explained by the forces of evil that dwelt inside the world.

Even the earliest scientists, men such as Aristotle, Thales and other philosophers of ancient Greece and Rome, couldn't escape the influence of these underworld myths. The classical philosophers believed that four fundamental elements—earth, fire, air, and water—made up all else in nature, and they looked to these elements to explain the natural disasters of their Mediterranean home. They tried to eliminate all supernatural explanations from their theories of the earth. Yet those theories were inevitably haunted by one recurring image from the mythologies of their day: a cavernous void inside the earth.

This void plagued the ideas of Anaxagoras, a Greek philosopher of the fifth century B.C. He believed that the element fire caused earthquakes—a conviction that came to him in a storm. Having seen two dark clouds come together to make lightning, he concluded that thick clouds of vapor in subterranean caverns also collided and belched fire. Rushing through an opening to the surface, the fire would shatter all obstacles in its path, sending shock waves thundering through the ground and lava streaming from the hole.

The same image of a hollow earth also cropped up in the writings of the Greek philosopher Archelaus. As an explanation for earthquakes, he proposed that air was gradually being forced into the earth's interior caverns through passages that extended to the surface. Once the caverns were filled, he said, any additional air compressed the rest, causing violent subterranean windstorms that were felt on the surface as tremors.

Aristotle, Democritus, Anaximenes and many others fell victim to the same delusion. In the first century A.D. the Roman philosopher Seneca summed up eloquently the image of the earth's interior that most of the classical philosophers shared:

> Be assured that there exists below everything that you see above. There are antres vast, immense recesses and vacant spaces, with mountains overhanging on either hand. There are yawning gulfs stretching down into the abyss which have often swallowed up cities that have fallen into them. These retreats are filled with air, for nowhere is there a vacuum in nature; through their ample spaces stretch marshes over which darkness ever broods.

Modern scientists marvel at this description. They know that the tremendous mass of the earth, crushed by gravity's unseen hand, could never accommodate the roomy caverns that Seneca envisioned. Nor can researchers imagine anyone walking down into the earth through the mouth of a volcano as Verne proposed several centuries later. Yet buried in these enduring myths is a picture of the global deeps that is strangely similar to what geophysicists are seeing today. Computers amassing the digital traces of thousands of seismic waves have revealed a buried landscape. Halfway to the center of the earth lies not a Central Sea but central continents with mountains and broad valleys. These deep features are constantly exposed to the cyclonic winds of molten iron whipping around the earth's core.

Today's scientists also hear the rumble of fire within the earth, though not as Anaxagoras envisioned it. They know that it isn't the crack of lightning from subterranean clouds or the roar of beasts within, but the sound of molten rock. No longer the stuff of myth, the earth's internal crackles and pops have become state-of-the-art tools. In recent years, they have given a new impetus to the forecasting of natural disasters.

Eruption of the Century

Miles below the visible part of a volcano, as magma begins to swell beyond its confines and carve tunnels into the surrounding walls, it generates a telltale sound known to volcanologists as a long-period event. The sound is caused by gas moving back and forth in the cracks of the earth, and it vaguely resembles the chime of a gong—an oscillating tone that resonates through the rock. Our senses can't detect these long-period events, but seismographs can. Volcanologists now recognize these oscillations as a red flag that a volcano is entering a dangerous phase. Early in June 1991, the Philippine volcano Mount Pinatubo started to ring.

Pinatubo had been showing signs of life for several months before that—enough for Philippine volcanologists to ask the U.S. Geological Survey to send down an advisory team. Once the team arrived, the signs quickly grew more urgent. First a dome of lava pushed its way through the volcano's flank on June 7, providing the first direct evidence that molten magma was coursing through this long-dormant volcano's veins. Shortly thereafter, long-period events began filling up the seismograms. That's when David Harlow, who headed the USGS's team of scientists, says he began to be concerned. In the days preceding the appearance of the lava dome, his team had done some geological reconnaissance work on the mountain, and everything they saw from past eruptions had convinced them that if this volcano was going to erupt, it was going to erupt on a massive scale.

In some respects, that made their job easier. The group of scientists from both the Philippines and the United States recommended evacuation. The challenge was to persuade government officials to act on that recommendation, impressing upon them that even though nothing spectacular seemed to be happening on the mountain, the potential for disaster was enormous. Several days of intense conversations culminated in the evacuation of about 200,000 Filipino residents and 15,000 American military personnel and their families from Clark Air Base, leaving only a skeleton security force of several hundred uniformed men and women, as well as the scientists.

Once the base was evacuated, the panic lifted. But it was replaced by the dread that Harlow and his team had made a mistake. What if the volcano didn't go off? How would they explain their actions to the American military and the Philippine government? The scientists were praying for an eruption. Then on the morning of June 12, their prayers were answered.

Against a crisp, blue sky Mount Pinatubo let loose an enormous eruption that blew thousands of tons of ash and dust ten miles into the sky. Fifteen miles to the east, at Clark, the skeleton crew, mouths agape, had a perfect view of this spectacular mushroom cloud unfurling.

There was no need to run; this eruption didn't put them in any danger. Over the next few days, though, the signs of what was happening underground became more severe. The explosions continued and increased in power, and the long-period events grew so intense that the scientists at the base could feel them rolling underfoot.

By the fifteenth, conditions were truly frightening. Burning rock came racing down the mountain in chunks, laying waste to everything in its path. A typhoon also moved in that day, obscuring the mountain and turning the airborne ash into a downpour of mud.

The geophysicists grabbed their gear and raced away from the base along with the handful of airmen. Then, in a moment of daring, the scientists turned around and went back. They'd decided that they hadn't shown the proper measure of fortitude. By the time they got to Clark, however, the sky was completely black. Mud and golf ball–sized chunks of lava rained down on them. Any foolhardiness they once had soon disappeared, and Harlow and his team proceeded to a fallback position where they could monitor the volcano in safety.

Nearly five hundred people were killed in the Pinatubo eruptions—most from buildings that collapsed under the weight of the sodden ash. But without the scientists' warnings, the death toll could have been in the hundreds of thousands. The scientists' quick action is just one example of how the field of geophysics has advanced over the last decade.

Prior to this, researchers had simply listened to the ground's rumblings and hoped they would see a clear increase that would warn them when the eruption was imminent. The situation was akin to guessing when an argument would turn into a fistfight based solely on the volume of the shouting. By studying many volcanoes, however, scientists are learning what the varied murmurings mean about what is happening underground. Listening to this subterranean activity has allowed them to make predictions about how and when each pool of magma will erupt. In the case of Pinatubo, their predictions were timely and of lifesaving consequence.

But not all of the advances over the last decade are directed toward potential cataclysms at the earth's surface. What Pinatubo's researchers did with the shallow seismic waves from this volcano is similar to what geophysicists are doing on many levels. Using indirect methods, they are beginning to form a picture of the deep earth, and thus to comprehend how the world behaves.

This is today's oracle at work. After centuries of piercing and probing the earth, geophysicists are bringing down from the mountain a deeper understanding of the planet and its upheavals. They are beginning to grasp the subtle language of its hidden motions. Consequently, they are telling an entirely new story of the earth.

Like the theories of the ancient philosophers, that story is based on observations of the natural world. But in this modern scientific era, geophysicists' powers of observation are extraordinarily advanced. Gone are the days when scientists were blind to the world beneath their feet. Gone are the days of "stamp collecting," as one researcher called the years spent crisscrossing the earth's surface, identifying bits of rock and cataloging where they lay.

Modern researchers have invented ways to peer into the earth just as if they were digging into it or slicing the planet in two. They have devised methods of mimicking the earth's extremes of pressure and temperature. Using powerful precision vises and controlled explosions, they can now re-create the conditions of the interior in their laboratories. And with these sophisticated new tools, geophysicists are tracking the forces that shape the face of our planet.

The last decade of research has thus been a journey to the center of the earth, not in fiction but in fact. In some cases the techniques for exploring the deep earth are so new that geophysicists are rushing to tackle questions almost as fast as they can think of them. For similar reasons, the handful of tools geophysicists have for sensing the planet are now stretched to the limits. These scientists cannot literally march through the bowels of the planet as Verne's protagonists did. What they see, they view from afar. As Florida State University's David Loper once put it, "All of us are in awe of the astronomers and the astrophysicists who are able to predict the composition of the stars. But in fact they have a pretty easy job. They can *see* what they're working on. In geophysics we have a little bit of a problem. There are about three thousand kilometers of rock between us and what we want to look at.

And we have to use all the ideas and observations that we have available to us to find out what's going on down there."

That sometimes means making educated guesses. Unable to see beneath a volcano, researchers can only theorize about what causes the pulsing tremors they hear within it. Similarly, they must use their imaginations to picture how rock currents stir the deepest reaches of the earth's mantle. For that reason the modern-day vision of the planet is still part science, part invention. In subtle ways, religion, history and popular culture may still be influencing how geophysicists see the earth, just as they influenced the ancient philosophers who wrestled with the same subject.

Yet that in no way detracts from the value of this endeavor. Today's geophysicists are assuming their place in the long line of philosophers and scientists who have tried to comprehend the world. Their role, like that of all the explorers before them, is simply to advance our knowledge of the earth on which we live.

This book is an attempt to capture some of that journey of discovery and to describe the emerging vision of the inner earth. Along the way it will take up the stories of a handful of researchers for whom the childhood wonder about the planet never ceased. Don Anderson is one such researcher—a Caltech geophysicist with a penchant for putting his latest ideas into verse. Years ago, I watched as Anderson chaired one of his last meetings as president of the American Geophysical Union, the professional society for this field. With a puckish grin and his trademark bolo tie, Anderson shunned the role of chairman and transformed the awards banquet into a kind of Quaker meeting. One by one, geophysicists stood up and said their piece, a toast or an observation, and then quickly gave up the floor.

The contribution of geophysicists to our understanding of

the earth is equally communal in spirit. Each researcher puts forth a piece of knowledge. He or she, in turn, listens to the discoveries made by others and tries to synthesize them into a coherent picture of the planet. In this way each researcher may carry away a working model for the earth that is richer and more detailed than the one he or she brought. This is the approach that has allowed geophysicists to progress over the last quarter of a century from the revolutionary theory of plate tectonics toward a new revolution—a theory of the whole earth and how it works.

Inevitably, these researchers differ in their views of the world. The latest findings are often met with contention, and as with any human endeavor, rivalries can cloud researchers' judgment in critiquing the findings of others. Where significant differences of opinion exist, I have tried to give the reader an idea of the strengths and weaknesses of each side's arguments. That is, after all, the safest route through the treacherous waters of a scientific conflict.

Yet even in light of the debates that mark this or any field of science, deep-earth geophysicists embody a common tradition of inquiry. Though they may possess a personal view of how the earth works, most are only too glad to yield the spotlight to the earth itself.

The earth is the main character of this book. Geophysicists' emerging views of the planet stand in stark contrast to the dull image of it as an inert ball of rock, good for dispensing oil and precious metals or entombing mountains of waste. Their descriptions of it brim with energy and tumult. Taking shape in their minds is a vivid new portrait of the inner landscape, rendered in the detail and depth that a decade ago could only be imagined. It is a world everywhere in motion.

THROUGH THE LOOKING GLASS

"That one'll be gone by tomorrow," the pilot's voice crackled over the headset. From a prop plane circling above Hawaii's southeastern shore, he pointed toward a small wooden house below. A red-hot finger of lava was slowly advancing toward the house, lapping the edge of the back lawn. From our vantage point in the air, this tiny patch of green stood out like an island in a sea of stone. All around the yard lay brittle black rock frozen in tortured positions.

We were flying over the former village of Kalapana, a lush seaside community steamrolled by lava in the spring of 1990. Under the gaze of thousands of tourists, geophysicists and civil defense workers, a swath of molten rock rolled through town, snaked across the highway that ran beneath Kalapana, and then made its final plunge into the sea. A white pillar of steam marked its finale.

This journey from the open sore on Kilauea's flank, where

deep magma burbles up to the surface, to the Hawaiian coast amounts to some seven miles. Normally lava streaming down the side of the mountain would have cooled and solidified long before it traveled this far. But in recent years the lava hasn't been taking this overland route. Great pipes of rock, wider than humans are tall and carved by the burning magma itself, wind beneath the island's slopes like giant water mains. They funnel what has lately been a continuous flow of the lava well down the mountain before the molten stream breaks out onto the surface.

Such a lava tube fed the Kalapana eruption, and now, as the prop plane circled through the thick sulfur mist, a last trickle of that flow was coming to claim one of the few houses left standing in the village. True to the pilot's prediction, within a couple of hours the molten rock had crossed the yard and reached the back porch. Like a sheet of crepe paper, the wooden house burst into flames and burned to the ground in minutes. Its lush quarter acre quickly disappeared into the charred surroundings.

Where once this corner of the island was swaddled in green, a lunar landscape now stands. The entire region has been cordoned off by civil-defense workers, making Kalapana a brand-new wasteland. For Hawaiians, however, the land is only temporarily lost. As the flows in this area let up, the road will be cut back through the lava. After a few decades dry scrub and, later, flowers will sprout upon the rock, just as they have throughout the rest of the island, where the land was laid down in exactly the same way. As Hawaiian Islanders have done for centuries, the residents of Kalapana will eventually rebuild their houses and farms and stores, erecting new villages on top of the rubble of the old.

Most of us live in neighborhoods far removed from the earth's internal fires. A river of magma doesn't run past our doorsteps or threaten the roads we take to work. But like the people of Hawaii, our communities are also perched atop a constantly shifting platform.

The surface of the earth is perpetually moving. Hills slowly rise to form mountains. Lakes deepen and oceans are born. Everything around us, from the Tien Shan Mountains in China to San Francisco Bay, has been forged by incremental changes and time. Though that time is measured in millions or even billions of years, the changes are detectable. In a single decade two houses on distant parts of the globe will move, ever so slightly, with respect to each other. They will drift an inch or two apart or shift their latitudes.

These motions aren't compromising property values, and they are far too subtle for us to grasp with our senses. But these changes are slowly, resolutely transforming the face of our world. To see them is to witness a monumental work in progress, an experience that geophysicists now enjoy thanks to the early advances of the space program.

Shooting the Moon

On July 20, 1969, Apollo 11 astronauts Neil Armstrong and Edwin "Buzz" Aldrin took their first, historic steps onto the surface of the moon. Bounding around the Sea of Tranquility in their oversized boots, the astronauts left more than their footprints on its powdery topsoil. They also left a reflector, positioned so that it looked back toward the earth. This reflector, a simple aluminum panel a foot and a half across, framed a hundred "corner cubes"—technologically advanced versions of the reflective discs that line the shoulders of interstate highways.

For scientists on earth the panel was a tiny mirror in which they could see the planet's reflection. By measuring the time a pulse of laser light takes to travel to the moon and back, researchers obtained their most accurate reading for the distance of that journey. But more to the point, if they measured it from two sites on the earth at the same time, they could learn the exact length of two legs of a triangle. Then they could calculate the third. That third leg would change as the earth's landmasses slowly drifted around its surface. Thus the moon mission had given geophysicists back home a way to track the world in motion.

Shooting the moon, they called it. When the lunar laser-ranging operation began, project scientist Eric Silverberg and his team at the McDonald Observatory in western Texas carried out a nightly target practice that would have been more at home in the celluloid frames of *Star Wars* than on those cattle-covered plains. The observatory's flickering green laser beam bisected the sky like a light-saber. But with a firing range of over 200,000 miles, hitting the panel on the moon was at least as difficult as shooting a hole in a dime two miles away. In the beginning they scored far more misses than hits. With time, though, the researchers learned how to aim their laser beam more accurately, as well as how to capture the weak, diffuse light that the target reflected. In the years since that first bull's-eye, in 1969, they have hit their mark thousands of times.

Their targets now include not one but four panels that rest upon the moon's surface, as well as a couple of oddly shaped reflectors in orbit around the earth. The *LAGEOS* satellite, set in space in 1976, looks like a two-foot-wide metal golf ball with a reflective disc in each of its dimples. Its quirky, spherical design was intended to minimize drag, making its orbit and therefore the measurements from it far more precise. In

October 1992, its sister satellite *LAGEOS II* went into orbit. With these and other laser targets, researchers can find the distance between any two sites, even if they are thousands of miles apart, to within two centimeters, or roughly the nearest inch.

They can shave that margin of error down even more with a similar technique called very long baseline interferometry, or VLBI. A hand-me-down from radio astronomy, VLBI uses the radio waves from distant quasars in much the same way that laser ranging uses light. The waves take slightly longer to travel to one receiver antenna than to a second somewhere else on the globe, and that delay is used to calculate the distance between them. Geophysicists can use VLBI to verify the measurements of laser ranging, and vice versa.

But the most promising addition to their toolbox is the Global Positioning System. The GPS has democratized the business of measuring crustal motion. Over the last twenty years the U.S. Department of Defense has put up a bank of twenty-four satellites to make it easier for soldiers to navigate unfamiliar terrain, such as the Iraqi desert. Portable GPS receivers carried by American troops during the Gulf War picked up signals from the satellites and triangulated between them to locate a soldier's position on the globe. In peacetime, geophysicists have been using the same receivers to track crustal movements, and they have found GPS to be cheaper and easier to use than VLBI and laser ranging.

These three techniques have brought the surface of the earth into focus. Hundreds of sites across the globe are continuously being monitored for their relative positions. And nothing emerges more clearly from these observations than the peripatetic nature of our crust. A site in Sussex, England, is drifting away from one in Greenbelt, Maryland, at a rate of about two-thirds of an inch a year. On the other side of the globe, two sites in Monument Park, California, and Simosato, Japan, are slowly converging.

Each of these incremental shifts represents only a brief moment in geological time. But when researchers look at a series of movements like the frames of a slow-motion movie, they see an unmistakable pattern. All of the earth's surficial motions can be divided among a handful of crustal fragments that resemble the shards of a broken eggshell. Seven major shards and several smaller ones glide across the earth's surface in a steady direction at a steady pace. They move about as fast as a fingernail grows—an inch or two a year. And upon these rigid platforms both the continents and the oceans ride side by side.

From the distant vantage point of space we can see the great North and South American Plates slipping westward away from the Eurasian and African Plates moving eastward. Across the globe, the giant Pacific Plate is drifting to the northwest and losing ground as the Eurasian Plate rides over it. Advancing toward this Eurasian-Pacific juncture is the Australian Plate, entering warmer climes. It is simultaneously pulling away from the great Antarctic Plate that seals off the southern end of the world.

The roving rafts of the earth's surface are known as the tectonic plates. Where they scrape against one another, earthquakes, volcanoes, mountain belts and all manner of geological phenomena arise. The theory of plate tectonics explains these phenomena and is commonly regarded as one of the cornerstones of modern geophysics. This theory also laid the foundation for the modern revolution in our understanding of the deepest parts of the earth.

In exploring the concepts behind plate tectonics, it is worth taking a look at how this initial revolution came about, for the theory was not always so well accepted. Nor was it always possible for geophysicists to see the motions of the surface so clearly. Plate tectonics had to be pieced together from many disparate lines of evidence.

That evidence came not from space but from the rocks around us. Throughout the world and throughout history, the disruptions that we now associate with plate interactions have taken place—from the eruption of Mount Vesuvius, which buried the city of Pompeii in A.D. 79, to the great San Francisco earthquake of 1906. For centuries people have scoured the solid ground in search of the causes of these catastrophic events. Their efforts fleshed out the modern field of continental geology.

Yet it was not until researchers extended the same effort to the oceans that the bigger tectonic picture fell into place. World Wars I and II brought tremendous gains in the technology for mapping the sea floors. As geologists put this technology to use, the true character of the earth's surface, both wet and dry, began to unfold.

The Plate Tectonics Revolution

One of the first to divine the secrets of plate tectonics was a quiet, devoted geologist named Harry Hess. Hess was a professor at Princeton University when the Japanese attacked Pearl Harbor on December 7, 1941. During his years of offshore research, he had also acquired the rank of lieutenant in the Navy reserves. The morning after the bombing, he reported for active duty. As part of his service, Hess joined the crew of the attack transport USS *Cape Johnson,* first as a navigator but eventually as its commander. From its decks he witnessed the landings on the Marianas, in the Lingayan Gulf and on Iwo Jima.

Between military missions, Hess's thoughts often turned to geology. Like most ships of its class, the *Cape Johnson* was equipped with an echo sounder, an instrument that sent out pulses of sound that bounced off the ocean floor and measured the time for them to return, thereby calculating the water's depth. As Hess voyaged around the globe in this ship, taking some liberties in how he carried out his travel orders, the steady ping of the echo sounder generated a profile of the ocean floor. Back and forth through the Pacific, Hess kept the echo sounder running day and night. Each path was only a single stroke, as in a pencil rubbing from a leaf. But combined, they sketched in a vivid image of that ocean's murky bottom.

Once the war was over, the mapping that Hess and others had begun continued at full force. Embroiled in the Cold War, the United States was eager to maintain its military strength, particularly in the competitive field of submarine warfare. In 1946 the U.S. Office of Naval Research was established. It provided extensive funding for ocean studies throughout the 1950s, and out of that investment came an explosion of information about the ocean floor. Like the blind receiving a miracle cure, geologists saw the submerged terrain materialize before their eyes.

The deep ocean floor is a vast landscape of broad, level plains. Kansas-flat basins the size of small continents are only rarely interrupted by an island or a seamount rising sharply from the horizon. For millions of years the debris of ocean life has sifted down through the water, burying the bumps and dimples of the bottom rock in a smooth layer of sediment.

Yet in the middle of these featureless plains snakes a striking range of peaks. Like the stitching on a baseball, these mountains line up one after another in a continuous chain. Some 40,000 miles of seamounts circumnavigate the globe. From a streak down the middle of the Atlantic Ocean, they cut around the southern tip of Africa, across the Indian Ocean and down beneath Australia, finally tapering off in the southeastern Pacific. These mountain belts are known as the mid-ocean ridges. Thousands of feet from the ocean floor the ridges rise and, unlike mountain chains on land, their highest peaks are split down the middle. Between the towering rows of spires lies a low valley up to a few miles wide. Pastures of fresh magma flows known as pillow lavas and the rubble of frequent earthquakes line the floor of this stone trough.

To postwar geologists, the origin of these mid-ocean ridges presented a tantalizing mystery. It attracted some of the fittest minds in geology, not the least of which was Hess's. In 1960 he attempted to explain these ridges along with a number of disparate observations about the ocean floor in an essay entitled "History of Ocean Basins." A brilliant scientist, Hess was unassuming about his talents and perhaps tentative about the new ideas he was presenting. In his introduction he adopted a cautious tone. "I shall consider this paper an essay in geopoetry," he wrote, hinting that his colleagues should keep an open mind about his conclusions.

Hess then proceeded to describe the ocean floor as if it were a collection of giant conveyor belts. At the beginning of each belt was a mid-ocean ridge, where new ocean crust was made. The rows of mountains at the ridge were actually symptoms of the stresses along these boundaries. Here the ocean floor literally split open as slabs of ocean crust tried to move, in their conveyor-belt fashion, away from the ridge. Hot magma from the mantle would rise to the surface to fill in the crack. The magma would then cool and harden, adding to the four-mile-thick slab of moving crust.

At the other end of the conveyor belt were the deep ocean trenches, plunging abysses over which geologists had puzzled for decades. Hess called them jaw crushers. Here, he believed, the ocean crust dived back into the earth to be destroyed. "The whole ocean is virtually swept clean, its rock replaced in the time it took to travel from the ridges to the trenches," he wrote.

The continents themselves were swept along passively by the ocean crust as it moved across the earth's surface. Built of lighter stuff than the ocean floor, the continental crust didn't sink down into the trenches when it reached this end of the conveyor belt. Instead, it came to a rest above the trenches. Like a bulldozer, it plowed rock off the ocean crust that *was* descending and piled it into hills along its edge. In addition, two continents could collide above a trench, Hess said, and then their borders would crumple like smashed fenders, driving up tremendous mountain belts along the land between them.

Hess's synthesis was a bold stroke of intuition. In this single paper he had captured the essential motion of the earth's crust, from its formation to its destruction. More than geopoetry, Hess's theory of seafloor spreading was the key to the earth's behavior. But most of his contemporaries didn't see it that way. At the time, the geological community, particularly in the United States, still believed strongly in an earth with an immobile crust. They greeted Hess's ideas with skepticism or outright disdain. They continued to pursue their own explanations for the mysterious mountains on the ocean floor.

The evidence proving Hess right wasn't to come until a few years later. This took the form of an unusual series of stripes, discovered as ships towing magnetometers trolled above the ocean floor. These stripes represented swaths of rock in which the magnetic field was normal, that is, aligned with the earth's present magnetic field, and swaths in which it was reversed.

To explain this mysterious zebra pattern, two English geologists, Fred Vine and Drummond Matthews, and Canadian geologist Lawrence Morley independently seized on the same approach. They combined Hess's seafloor-spreading theory with another, relatively new idea: that the earth's magnetic field had sometimes been reversed in the past. If it was true that the magnetic field flipped back and forth, then any new crust formed during a reversed era should reflect that. Whenever rocks are heated they lose any magnetization they might have had. Then as they cool, they adopt the orientation of the field around them. Vine, Matthews and Morley all believed that new crust forming from molten rock at the mid-ocean ridge would freeze the magnetic field of the moment into the ocean floor. That crust would then move away from the ridge, and when the earth's field reversed its poles, the newly forming crust would adopt the opposite magnetization. Over millions of years a pattern of alternating stripes, parallel with the ridge, would cover the ocean floor.

Upon publication, this theory of magnetic striping also received a cool appraisal. It was damned not only by its association with Hess's theory but by the complex and still unestablished concept of geomagnetic reversals. Yet when the Vine-Matthews-Morley theory was eventually put to the test, it passed with ease. Oceanographers turned up a series of ridges that fulfilled its predictions, beginning in 1965 with the Juan de Fuca Ridge off the coast of Vancouver. The stripes on either side of this ridge were both parallel to it and symmetrical about it, as if one side were the mirror image of the other. And many of the stripes could be matched up to specific periods in the earth's history when the magnetic field was thought to be normal or reversed.

At the same time that these stripes around the Juan de Fuca Ridge were vindicating Hess's ideas, the Canadian geologist J. Tuzo Wilson, working in the same region, seized on another piece of the crustal puzzle. A bold and enthusiastic researcher, Wilson was one of the few to have sided with Hess's theories right from the start, although his support did little to convince others. Wilson had only recently been an avid proponent of a theory that described the earth's surface as expanding, and before that, of a theory in which it was contracting. As a result, what credit Wilson received for his powerful intellect had been eroded by his scientific fickleness. Yet this time he was finally on the right track.

Examining the Juan de Fuca Ridge, Wilson discerned a third type of border in the conveyor belt of the earth's surface, a margin in which the crust was neither created nor destroyed. He described such margins as zones of transformation from one type of plate border to another, and he immediately dubbed them transform faults. Sections of the crust slide past each other along transform faults. As their rough edges try to scrape by in opposite directions, they move in fits and starts, releasing great earthquakes with each lunge forward. Such is the case with the San Andreas Fault. This deep groove through California marks the boundary where the Pacific Ocean floor is sliding by North America, and its seismic shocks are legendary.

By connecting transform faults with the theory of seafloor spreading, Wilson had made the link between all the active regions of the earth's crust: mid-ocean ridges, ocean trenches, earthquake zones and mountain belts. "These features are not isolated," he wrote in 1965; ". . . few come to dead ends. . . . They are connected into a continuous network of mobile belts about the earth which divide the surface into several large rigid plates."

With this insight Wilson had arrived at the essence of plate tectonics. Where in their travels the plates slam and scrape against each other, we find geological unrest. We find the Himalayas, products of a collision between the Indian and Eurasian Plates some 25 million years ago. We find the volcanic islands of the outer Pacific's Ring of Fire, a semicircle that extends from Alaska, through Japan and the Philippines to New Guinea and New Zealand. The islands in this ring spring from the descending Pacific Plate. As it has plunged into the earth, some of its rock has been softened by the earth's internal heat and has risen back toward the surface, poking neat arcs of volcanoes into the overriding plate.

From such rows of volcanoes, earthquake zones and mountain belts, geologists quickly carved up the world's surface. Soon the work that sprang from Hess's and Wilson's ideas accelerated to a feverish pace. While oceanographers tallied up ridge after ridge that fit the seafloor-spreading model, other scientists worked out the relative motions of the plates, finally showing in 1967 that what Wilson had conceived on a flat plane also worked on the spherical surface of the earth. The meetings that year were abuzz with excitement; geophysicists greeted one another in the corridors with the question, "Do you believe in plate tectonics?" Nearly everyone, it seemed, was in agreement. Plate tectonics was the simple, elegant theory that researchers had long been seeking.

The theory shattered many of the dusty notions that scientists had held about the earth. They had thought of the planet's surface as immobile and unbroken. That idea had its peak in the first half of this century in the baked-apple theory of the earth, which compared its wrinkled, pitted surface to the skin of an apple that had been baked and allowed to cool. Researchers believed that when the nascent earth had cooled and shrunk, ocean basins formed where the ground sank and continents formed where the ground remained high. But with tectonics, geophysicists progressed from thinking about continents and oceans as the enduring fabric of the earth's surface to a vision of everything adrift aboard enormous platforms of rock. At last they possessed a cogent framework for understanding their landscape.

That framework has held together for more than twenty-five years. It has endured the scrutiny of modern satellites and other measurement techniques. Indeed, the fact that the surface is broken into moving pieces is one of the few facts that a majority of adults know about the physical world. Perhaps the image is just so startling that it sticks in our minds. After all, the idea that the ground is in motion beneath our feet defies the senses. And like the realization that the sun doesn't really rise and set but remains still as the earth rotates, the theory of plate tectonics put an end to our naive faith in what our instincts tell us about the world.

2

JOURNEY TO THE CENTER OF THE EARTH

Many of the geophysicists working in the field today were in their late teens and twenties during the years when plate tectonics was sweeping earth science off its feet. It was the excitement surrounding this new theory that drew many of them into the fold.

Al Duba, who now runs a high-pressure-geophysics lab at the Lawrence Livermore National Laboratory, vividly remembers his first experience with the emerging theory. In 1968, Duba was at the University of Chicago taking a course with Peter Wyllie, a geophysicist who later wrote the standard college textbook on plate tectonics. But at the time, tectonics was still breaking news, and Wyllie, like many earth-science professors across the country, asked his students to debate the new theory.

To Duba and his team fell the unfortunate task of arguing

against plate tectonics. Yet even though Vine and Matthews's paper providing a critical piece of the theory had already been published, the evidence was apparently fresh enough—and Duba's team sharp enough—that when the faculty and students voted which side they believed, plate tectonics lost! Duba recalls that even though his heady victory was a mistaken one, it launched him on a long and successful career in geophysics.

Duba wasn't particularly interested in rocks before then (though he says the fact that his father was a coal miner may have influenced him). He went to the University of Chicago to study the physics of electricity. But having been awakened to earth sciences by the plate-tectonics revolution, he soon fell in with a professor who was squeezing minerals in order to understand the extremely high-pressure conditions inside the earth.

Many geophysicists of the deep earth have similar tales. Like Duba, these researchers drifted into earth science from fields like engineering, mathematics or computer science. Lured by the excitement of plate tectonics, some felt that earth science would be a practical way to apply their knowledge. Others were just looking for a profession that would allow them to work outdoors. But most came under the tutelage of a handful of researchers who had been thinking about the earth's interior for some time. This group of mentors had laid the foundation for deep-earth studies. They could map out the intellectual problems that faced young researchers like Duba, and impress upon them the excitement that lay ahead now that plate tectonics had given the field such a boost.

For once the theory of plate tectonics took shape, a series of implications for the deep earth quickly tumbled into place. If the global surface was in motion, geophysicists realized that the rock beneath it must also be stirring. Otherwise the steady advance of the plates would have ground to a halt millions of years ago.

This notion was not entirely new. Decades before, a few researchers had proposed that the inner earth might be in motion. Stoked by heat, they said, the thick layer of rock known as the mantle could be supple enough to flow. But at the time, these ideas were roundly rejected by geophysicists. Once again it was a case of letting their preconceptions about the world get in the way. Scientists knew that the mantle was made of solid rock, and the suggestion that a solid could flow like a liquid was considered absurd.

Only years later, when the theory of plate tectonics emerged, did most geophysicists change their minds. By that time science had advanced to the stage where researchers understood that even solids can, in some cases, flow like liquids. For example, the glass in the ancient windowpanes of medieval churches flows even though it is a solid. The windows are all slightly thicker at the bottom edge; for many centuries the glass has moved under the force of gravity. The mantle has approximately the same viscosity as glass, which means that the notion of a moving mantle is not unthinkable. By the late 1960s, researchers understood that the tremendous heat and pressure inside the earth, sustained for millions of years, could generate powerful currents of rock.

To understand these currents, geophysicists looked to the behavior of ordinary fluids. They began to think of the rock in the mantle churning in much the same way that soup does as it's heated on a stove. That soup will rise in some areas, spread across the surface, and then sink back to the bottom in other spots, forming continuous loops. Scientists call this sort of mixing convection, and it is readily apparent in the patterns it creates on the surface of a liquid. Left unstirred, simmering soup will produce a scum that gets caught up in the eddies.

The scum is brushed aside by rising currents and then, too light to sink, it collects above downwellings. Researchers thought that, like this scum, the plates might be moving with the mantle's eddies. The rock below could be rising to the surface at mid-ocean ridges and then sinking back into the interior along the ocean trenches. The plates would then presumably move along with the currents, connecting the upwellings and downwellings in giant convection cells.

In hindsight, these ideas about mixing seem fairly simplistic to modern geophysicists. The discoveries of the last decade have shown that a real understanding of mantle convection was still far beyond the early researchers' grasp. Yet for this small but growing group of deep-earth geophysicists, the uncharted underworld presented a rich store of possibilities.

Here was a brand-new kind of earth—a roiling, boiling dynamo. The ground was merely a collection of rafts caught up in its currents of stone. What did they look like, these mysterious rock currents that powered such movement on the surface? Were they tearing cracks in the crust? Were they shoving the surface plates around or making the continents bob up and down? Even at the time, researchers realized that their simplistic image of the inner earth could not explain everything they were seeing on the surface. For in addition to roving plates, geophysicists had also identified some aspects of the earth's crust that didn't fit into their neat tectonic scheme.

Picture the Hawaiian Islands poking out from the glistening blue expanse of the Pacific. With all that researchers knew about the power of tectonic motion, these lava mounds remained a mystery. Smack in the center of the giant Pacific Plate, they were far removed from any plate boundary that might have produced them. Throughout the world, researchers found similar island chains, isolated yet volcanically active.

How were these perfect lines of islands carved into the moving plates?

One day, when the ideas of plate tectonics were just beginning to brew in his head, J. Tuzo Wilson was flying over Hawaii looking down on the chain of islands below. It struck him that something could be lying beneath Hawaii, punching holes into the Pacific Ocean crust much the way a blowtorch could burn holes through a sheet of tin. Such a "hot spot," Wilson envisioned, would stay in one place as the plate drifted over it. He believed it must have built not only the Hawaiian Islands, but also the continuous trail of mostly submerged seamounts extending all the way back to the ocean trench that borders Alaska's Aleutian Islands.

Wilson then proposed that hot spots all over the globe possessed this same steadiness. They weren't drifting along with the crosswinds of plate motion but were instead behaving like stationary spigots of fire. That meant that somehow the spouts of magma that lay beneath these hot-spot volcanoes had to persist in spite of the mantle's steady churning, like surreal columns of factory smoke that never wavered in the earth's breezes. This mysterious notion of hot spots added yet another layer of intrigue to the inner earth's movements.

Stoking geophysicists' interest even further were clues about the enormous stores of energy in the earth's interior. Those clues came from a source even deeper than the mantle. Just as the brightness of a twinkling star gives away the fires of fusion burning inside, so the cloak of magnetism that hovers at the surface of the earth gave away the heat and tumult that lay within its core.

By the late 1960s, researchers already knew that whatever was causing the earth's magnetic field was not like the simple magnets that lie on a table and radiate magnetism. Instead, geophysicists believed that it was more like an engine. To work—that is, to produce the earth's magnetic field and keep it going continuously—this engine needed a supply of energy, and a simple calculation showed that the energy it needed was enormous. Somewhere inside the earth was a source of raw power that no one understood. To geophysicists this was a seductive notion.

Indeed, all the hints of a dynamic underworld lured geophysicists to studies of the deep earth like the smell of a sumptuous banquet draws a gastronome. They presented a challenge that researchers could sink their teeth into—intriguing, unanswered questions that could fill entire careers. Geophysicists were no longer thinking about just explaining the features at the surface. Volcanoes, tectonic plates and magnetic fields would make up a large part of the picture, but researchers saw an even bigger nut to crack. This broader problem touched on such philosophical conundrums as who we are and what our place is in the universe. Geophysicists were looking to tackle the question of how the whole earth works.

Just to begin to penetrate the mystery of the deep earth would take geophysicists above and beyond the current reach of their science. The dynamics of the interior would demand an understanding of complex, three-dimensional mixing, and farsighted geophysicists recognized that this was to be the new theoretical frontier. Having taken up the challenge, they now faced a great hurdle—finding ways to see beneath their feet into a world that was largely unexplored.

Digging Deep

Ironically, one of the most ambitious efforts to see into the earth was taking a nosedive just as the theory of plate tectonics was taking hold. In the middle of this century, earth scientists were casting around for a project that would take their field a giant step forward. Ambitious scientific efforts such as the space program had captured the public's imagination, and these researchers wanted something equally dramatic—their own version of the space probe.

At that time a loosely based, somewhat whimsical organization of scientists existed under the name of AMSOC, short for the American Miscellaneous Society. AMSOC had no president, nor even a list of members, though it counted some of the most esteemed American scientists among its ranks. They reveled in AMSOC's open aversion to the stuffiness and bureaucracy of other scientific societies. Meetings took place in bars or anywhere that two or more members happened to be present. Out of these ad hoc sessions came a bold plan to drill a hole into the earth.

This was to be no ordinary hole. Drillers would plunge through a record depth of water to pierce farther into the ocean floor than anyone had gone before. The trophy of the mission lay waiting for them at the bottom of that hole: a pristine piece of the earth's mantle. No one had ever seen a sample of this enormous layer of rock. Though scientists had hunches about what it would look like, they believed the hole in the bottom of the sea would answer their questions once and for all.

They also knew that such a scientific coup could only be accomplished by drilling into the seafloor, for the continental crust was simply too thick. Compared to the enormous depth of the globe, that thickness is trivial—a mere sliver on the surface. But the trip from a continent's grassy plains to the first layer of mantle rock would have been a journey all its own.

Sink a spade into almost any patch of ground and it will turn up a chunk of rich, brown dirt. Millions of years of plant and animal detritus have built up this loamy mass, tens to hundreds of yards thick at its deepest points. Where glaciers have scraped it off or rain has eroded it, the topsoil is thin. Yet even in these spots a spade still has to pass through rusty layers of clay, pebbles, and then fist-sized rocks before finally finding solid bedrock.

The bedrock extends down even farther, each layer overlapping another. Some of the rock has forced its way up into other layers like an errant thumb. Some of it has been sliced off a base of rock thousands of miles away and transported by tectonic forces. All the folding and colliding the crust has done throughout the eons has shuffled the different types of stone like a deck of cards. Limestone, granite and shale—sedimentary, igneous and metamorphic rocks—lie one atop the other in a mountain of colored strata.

These horizontal layers are readily apparent wherever a road cuts through the side of a hill, but by far the most glorious view of the bedrock is in the Grand Canyon. Ribbons of red, green and ocher stone stretch across the canyon walls. As vacationers hike down into the canyon or raft along the Colorado River, they can follow the layers of rock ever farther into the earth. At the deepest part of the canyon, beneath the hodgepodge of layers, can be found basement rock—the stable heart of the continent. Mostly granite and metamorphic rock, it has been heated, stirred and compressed for millions and sometimes billions of years. It forms a landscape of unceasing gray when viewed from the chasm known as Granite Gorge, one of the rare places on earth where this rock is exposed to the surface.

Geophysicists believe the basement rock extends for tens of miles into the earth. Only then comes the point where the crust finally ends and the mantle rock begins. That point is marked by a sharp boundary known as the Mohorovičić discontinuity, or Moho. It was discovered by the Croatian seismologist Andrija Mohorovičić in 1909 after he noticed that the tremors from a strong earthquake near Zagreb, Croatia, bounced off it and headed back toward the surface.

Since Mohorovičić's day, this spherical boundary has been mapped all across the globe. Beneath the continents the Moho lies anywhere from twenty to forty-five miles deep, much farther than anyone could dig. But beneath the oceans the layer of crust has only recently been formed by the mid-ocean ridge and then dusted with sediment. Because the ocean crust is so thin, the mantle is only a few miles away; hence the idea of digging beneath the churning waters of the ocean, through the hard volcanic ocean crust, to snag a piece of the elusive mantle—a feat that quickly came to be known as the Mohole Project.

It started off with a bang. Soon after its conception, the project began with a pair of test drillings in 1961, first off the coast of California and then in the deeper, rougher waters near Guadalupe Island, 250 miles south of San Diego. Researchers had stuck a drill rig onto a freight barge with a hole chopped into its middle and produced the scrappy vessel *CUSS 1.* With it they would attempt a novel sort of drilling. Using four outboard motors, two on each side, the helmsmen would have to keep the barge from moving more than a ship's length off its mark as the drill string plunged through the ocean depths.

By the time the second test drilling came around, the Mohole Project had caught the attention of the editors at *Life* magazine. They flew out author and sometime oceanographer John Steinbeck to chronicle the excitement on board the *CUSS 1.* His daily log, published under the virile title "High Drama of Bold Thrust Through Ocean Floor," accompanied a glorious photo essay, and Steinbeck, in his usual fashion, captured all the romance and passion of the mission from the "heaving and pitching" of the deck to the balletic grace of the drill crew, who had to time their movements with precision lest they be knocked senseless by a length of steel pipe gone haywire.

For two days, working twelve-hour shifts, the drillers sent the pipe down section by sixty-foot section, finally touching bottom around noon on the second day. The first core came up full of gray-green silt, sand and clay—the fluffy carpet of the ocean bottom. Bolstered by their success, the drillers continued on toward the next layer, the hard igneous rock of the ocean crust that geophysicists call basalt. But an attempt to pull up a second core failed when the corer snagged and then broke off its cable. The drillers had to start all over. Twenty hours dragged by as they pulled up the drill string and dropped it back down again. All the while the threat of defeat tainted the air.

Then on Easter Sunday, in what must have seemed like divine intervention, the drillers broke through. They succeeded in reaching the second layer of ocean crust, never before seen, and brought up a core of deep-blue, veined basalt. It was as historic a piece of the earth's crust as the Rock of Gibraltar. All those on board were yearning for a fragment, including Steinbeck, but this core was too valuable for the researchers to allow such souvenirs. (Steinbeck confessed to having stolen a tiny sliver, but then the chief scientist smuggled a piece to him, making the author so miserable that he had to put the first one back.)

With this successful test drilling, the Mohole Project was off to a heady beginning. The scientists had drilled nearly 600 feet into the ocean bottom in over ten thousand feet of water. Nevertheless, they still had a long way to go. The real Mohole, slated for a site off the coast of Hawaii, would be drilled in fourteen thousand feet of water through seventeen thousand feet of rock, or some three miles. Scientists set out to accomplish the next phase, only to find that it did not proceed as smoothly as the first. In the end, the Easter breakthrough marked the climax of the Mohole Project.

One by one, the governmental and scientific groups in charge of the project began to distance themselves from it. Its cost, vastness and complexity were too much to handle with the sort of ad hoc administration that saw the first phase to its goals. The National Science Foundation, which was footing the bill, decided to hire an independent contractor to complete the project. This step brought accusations of political patronage. At the same time there were growing disagreements over the scope and objectives of the project. Scientists and administrators bickered over whether this should be a continuing program in ocean-bottom drilling or a one-shot drive to the mantle. Year after year, the price tag for the project swelled, at a time when the costs of the Vietnam War were also on congressmen's minds. In the mid-sixties, *Newsweek* ran a story entitled "Project No Hole?" and indeed it was the beginning of the end. The haggling among scientists continued, and seeing the project in disarray, Congress eventually cut off financing for the Mohole Project.

Congress did, however, keep a trickle of funds flowing for shallow ocean drilling, and over the years numerous holes a few hundred yards deep have been drilled in waters across the globe. In an unusual postscript to the aborted Mohole mission, ocean drillers finally succeeded in snagging the first piece of the shallow mantle in January 1993. They dug in ocean crust where part of the mantle had been shoved close to the surface by the faulting of the seafloor. Though the researchers didn't actually reach the Moho, the piece they retrieved once lay near that boundary. Someday geophysicists may pierce that elusive part of the earth.

But the true failure of man-made holes, at least where deep-earth researchers are concerned, is that they require so much effort and money to dig without providing equal gains in return. Judging by the progress in drilling on land over the last century, one geophysicist estimated that, at most, it might be possible to drill a hole 35 miles deep by the year 2100. But the entire mantle extends as deep as 1,800 miles. Thus, in terms of what researchers need to understand the earth's interior, such a hole would barely scratch the surface.

By the early 1970s, this realization had already hit home. The small group of deep-earth geophysicists, fresh from their disappointing experience with the Mohole Project, understood clearly that no hole dug by man would open a window onto the churning earth. Fortunately, nature has been kind enough to dig a handful of holes herself. These holes have been much more helpful to deep-earth geophysicists over the last two decades. Samples of rock from them have provided researchers with most of their knowledge of the chemical makeup of the mantle.

Before researchers could exploit the riches of the holes, though, they had to find them. Over a century ago, a series of accidental discoveries made this insight possible. For their glimpse into the earth, modern researchers are indebted to the age-old allure of diamonds.

Diamond Pipes

In 1866, the first South African diamond was unearthed, and like the discovery of gold in California, the promise of instant wealth turned thousands of Africans into prospectors. They staked their claims along South Africa's major rivers, the Vaal and the Orange, since all the diamonds that had been unearthed anywhere in the world up to that point had come from the gravel of riverbeds.

Far from these rivers, the foreman of a farm in Jagersfontein, South Africa, had also caught diamond fever. This foreman, named Jaap de Klerk, had learned the rudiments of diamond prospecting from the many diggers who passed through his farm on the way to their claims, and in 1870 de Klerk set out to try his own luck. With the simplest of tools and a homemade trough, he began sifting through gravel from the farm. In less than a month he had turned up a fifty-carat diamond.

Naturally, de Klerk's hefty diamond struck his fellow prospectors as something of a fluke, since he had found it at some distance from a river. That and the relatively high price that the farm's owner was charging for digging rights discouraged most prospectors from staking a claim at Jagersfontein. But over the years a few prospectors kept up the search on the farm, and their efforts were richly rewarded. Jagersfontein seemed to possess a peculiarly large supply of high-quality diamonds. As the diggers kept looking for ever more concentrated sites, they gradually found themselves being drawn toward a particular section of the farm. By 1878 they realized that this region was the source—the rock in which the diamonds originated before being washed away by rains and floods. The beautiful blue-black rock that encased the diamonds formed an enormous cylindrical seam that the prospectors could follow deep into the ground, as if they were digging their way down the mouth of a pipe.

In fact, the Jagersfontein miners had stumbled onto what is now known as a kimberlite pipe, a unique type of fossil volcano that erupted roughly a hundred million years ago. Kimberlite pipes are the source of all the world's natural diamonds. And because of de Klerk's chance discovery, Jagersfontein was the first place where this connection between diamonds and the bluish kimberlite rock was made.

These pipes, which are scattered throughout the world, are the sites of some of the most powerful and explosive blasts in geological history. Scientists know that the ancient eruptions had to have been violent because the molten kimberlite, forcing its way to the surface, not only carved an enormous pipe (which in the case of Jagersfontein covered two dozen acres), but in addition brought up diamonds intact. These gems were forged from carbon at over 125 miles' depth, where the heat and pressure are great enough to form strong chemical bonds. Released from those conditions, diamonds will revert within minutes to the more pedestrian substance graphite. But in a kimberlite pipe the journey to the surface was so fast, taking minutes or even seconds, that the diamonds didn't have a chance to revert. In a flash they were cooled and decompressed to the conditions on the earth's surface.

For dozens of years miners have gone after these valuable gems. They originally mined the kimberlite pipes as open pits and then, when the diamonds on the surface were exhausted, mining companies began installing elevators along the sides of many pipes. Miners now travel over a quarter of a mile down through the earth in metal cages, into the rising heat of the deeper rock. Rock dust and the damp, sweaty smell of bodies fill the air as the African laborers, under the constant threat of a cave-in, carve their way into the diamond pipe from the sides.

Geophysicists are particularly indebted to these miners, though for reasons that have nothing to do with diamonds. In addition to precious gems, kimberlite also forces another harder type of stone to the surface: xenoliths, from the Greek for "foreign" and "rock." Xenoliths have always been mined alongside the kimberlite, but until recently their value was unrecognized.

At the turn of the century, when Jagersfontein was in its heyday, miners spread the kimberlite out on areas the size of football fields, poured water on it, and allowed the elements to take their toll. Then the men sifted through the material and concentrated the diamonds. Since they knew there were no diamonds in these harder rocks, they just left them out in the fields.

Nowadays, deep-earth geophysicists spend weeks at a time scouring these fields in the searing South African sun in order to find xenoliths. These exotic rocks were dragged up the pipe, sometimes ripped off its walls, as the molten kimberlite raced to the surface. Thus, such rocks originate at depths much greater than any man-made hole has ever reached.

In fact, no one has ever beheld a rock from deeper in the earth. Xenoliths are the only firsthand information that researchers have on what lies beneath the crust and are therefore the only direct clue to the makeup of the mantle. By looking at xenoliths that came from about a hundred miles down, researchers discovered sometime ago that the main component of the shallow mantle is a particular olive-green rock known as olivine. An entire beach of olivine, known as Green Sand Beach, lies at the southern tip of Hawaii, where anyone who is curious can go and run his or her fingers through a handful.

In 1990, researchers took an even deeper step into the mantle. On the rock fields of Jagersfontein, Stephen Haggerty, from the University of Massachusetts, and Violaine Sautter, from the University of Paris–South, found the first "ultradeep" xenolith. The researchers proved that this rock came from a depth of at least 250 miles—roughly doubling the record of the nearest contender. And in the years since, researchers have discovered in diamonds from a Brazilian mine a series of mineral specks that they believe to be samples of the mantle at more than 400 miles' depth. An important transition zone exists in the mantle between 250 and 400 miles. As a result, both sets of deep rocks have fueled an ongoing debate over how this transition zone looks and why it exists.

We will leave the details of that debate until a later chapter, for 250 miles is still only a fraction of the way into the earth. While the ancient blasts that punched out kimberlite pipes helped researchers see into the earth farther than before, these holes in the ground, or any dug by man, couldn't take geophysicists as deep as they needed to go.

Researchers recognized that they simply couldn't study the interior the way they had studied the surface for so many decades. They would never be able to examine the rocks of the deep earth for signs of motion. Nor could they dip their instruments into boreholes the way they might dip their hands into a stream to gauge the speed of the current. The rock currents in the mantle were running far too slowly for geophysicists to measure directly.

In effect, researchers were denied the direct use of their senses. But a blind person can still recognize a friend by the sound of his footsteps or even his scent. Losing some of their most trusted tools encouraged geophysicists to develop others. For there is more than one way to see the inside of a planet.

3

CAT-SCANNING THE PLANET

In the late 1970s, geophysicists began to retire the old analog seismographs that had documented the earth's tremors for generations. The tiny pens, scrawling in palsied traces on endless white ribbons of paper, slowly ground to a halt. In seismic stations throughout the world, modern digital seismometers took their place.

These new machines could pick up much fainter shudders from earthquakes around the globe. But more important, they stored their data in the digital form that computers so readily digest. That advance brought geophysicists out from under a mountain of paperwork. Before these instruments came online, researchers used to travel to the relatively few sites where seismograms were stored on microfilm and make photocopies of the archived records. Then, with thousands of pieces of paper in hand, they would go back to their offices and, one by one, trace over the squiggly lines on an electrical table to enter the information into a computer.

But with digital instruments and digital storage, the data could be transferred through phone lines from the source to the computer. Researchers could scan tens of thousands of seismograms in a week instead of months. The change in instrumentation thus touched off an explosion of information.

As of 1977, fewer than ten digital seismometers were deployed around the globe. By 1983, that number had grown to fifty. Tremor by tremor, the machines had amassed an enormous pile of data that more than one seismologist felt was ripe for harvest. From opposite coasts of the United States, two groups of researchers converged on the data set simultaneously, eager to see if it would fulfill their expectations. They were looking to these waves for signs of motion, hoping to transform science's static image of the world into one of suspended animation. Competing against time and against one another, they tweaked and massaged the raw seismic data with complex computer models. In retrospect, they were spinning straw into gold.

Windows onto the World

Earthquakes have always been our finest windows onto the world. Their tremors pass through the planet like X rays through flesh and bone, bringing us clues about the rock they penetrate.

In the United States, most of those earthquakes take place in California, where the land is riddled with faults. Along the granddaddy of those faults, the San Andreas, strain from the movements of the Pacific and North American Plates has been building up for so long that residents are bracing themselves for "The Big One," an earthquake of magnitude 8 or more. In the meantime, they have endured a series of lesser but still deadly ones, such as the magnitude 6.7 earthquake that struck the town of Northridge, twenty miles northwest of downtown Los Angeles, on January 17, 1994.

This violent quake shook the L.A. basin like a cocktail. The main shock, at 4:31 A.M., jolted millions of residents out of bed and sent them onto the street. There they found that, not only had their electricity been cut off, but gas and water mains had ruptured as well, causing fires several stories high to burn on flooded streets. Roving camera crews quickly transformed the wholesale destruction caused by the tremors into a series of images that played across television screens nationwide: the footage of an apartment building in Northridge that had collapsed from three stories to two; a fallen section of the Golden State Freeway that took the life of an off-duty police officer; and local citizens working side by side with emergency crews to pull their neighbors out of the rubble. In all, fifty-seven people died and at least $15 billion in property damage occurred as a result of this earthquake, making it overwhelmingly clear how vulnerable this populous city is to shaking.

Yet the Northridge earthquake was relatively small. For every increase of one on the Richter magnitude scale, the ground motion jumps tenfold and the amount of energy released by the earth rises about thirtyfold. L.A.'s recent scenes of damage brought back memories of an even more powerful and deadlier California quake—the magnitude 7.1 jolt on October 17, 1989, that interrupted the San Francisco World Series.

Known to scientists as the Loma Prieta earthquake, it emerged from a spot eleven miles beneath the surface of Loma

Prieta, the highest summit in the rugged Santa Cruz Mountains. Sixty miles from San Francisco, two rock faces of a fault suddenly lurched past each other, and like rolling thunder, tremors from the jolt swept north past San Jose and Palo Alto into the heart of San Francisco.

These tremors were similar to those that emanated from Northridge. In fact, they are similar to the seismic waves from any other earthquake anywhere in the world. But the fact that the Loma Prieta earthquake was so strong meant that those who witnessed it may have been able to detect a distinct beginning, middle and end to the experience. These stages are a result of the different batches of seismic waves that such a jolt releases. Each type of wave has a character all its own and each provides geophysicists with a different kind of information.

The first seismic waves to strike San Francisco in 1989 were the P waves, which produced the low rumble that presaged the violent shaking to come. *P* stands for *prima,* Latin for first, but geophysicists often refer to these as pressure waves, because as the P waves travel through the earth, they compress and then expand the rock along their paths. The effect is something like that of stretching a Slinky taut, squeezing a few of its coils between one's fingers, and then letting go so that the wave of compression travels down the length of the spring. But instead of the tinny rippling of coils, San Franciscans heard the throaty roar of rock.

Some 58,000 baseball fans who had crowded into Candlestick Park and were waiting for the third game of the World Series to begin heard this low rumble at just after five o'clock. Earthquakes are such a common topic of conversation in California that nearly everyone in the stands must have known what was coming next. Yet there wasn't time to run for cover.

The fans braced for the seismic waves that would soon have the ground beneath them rolling like the surface of an ocean.

True to form, the stadium seats beneath them started to jerk back and forth and up and down just a couple of seconds after the first rumble. These were what scientists call S waves, *S* standing for *secunda,* or second. They are fast, short waves. Small buildings, like small boats on a choppy sea, are rocked especially hard by their passing.

Then, seconds behind them, came the longer waves, the Love and Rayleigh waves, named after the British mathematicians who discovered them. Like giant swells rocking an ocean liner, these waves shook larger objects. Riding on Love waves, the city's skyscrapers yawed from side to side, and Rayleigh waves raised and lowered the buildings in a regular, elliptical pattern like the motion a piece of driftwood makes as it bobs along the ocean's surface.

In fifteen seconds the strongest shake was over, and surprisingly no one at the stadium was seriously injured. A startled cheer erupted from the crowd. They had survived what was, at least, a Pretty Big One—the largest earthquake to strike northern California in eighty-three years.

From the stadium, however, the Loma Prieta seismic waves continued on past San Francisco to all corners of the globe, squeezing and jiggling the rock as they went by. The Love and Rayleigh waves followed paths that skimmed along the curved surface of the earth, dipping only into the uppermost part of the mantle. Geophysicists refer to them as surface waves. But the P and S, or body, waves traveled through both the deep and shallow layers of the earth. They fanned out in every direction from the source of the quake, reaching the opposite side of the earth some twenty minutes after the initial shaking. Everywhere on earth, researchers knew that a quake had just occurred.

A Seismic Legacy

Since the late 1800s, scientists have been recording the tremors from distant earthquakes at seismic stations throughout the world. In any given year the number of quakes, both large and small, is in the thousands. They emerge not just from California but from fault-riddled crust all over the globe. And every time one rattles the surface, geophysicists learn a bit more about what's going on inside our world.

Each earthquake gives them a fleeting glimpse into the planet's interior, like a flashbulb lighting up a darkened room. Sometimes geophysicists even refer to seismic waves as "lamps" because of the many similarities they have with waves of light. Like light, seismic waves can reflect and refract off a surface. Shafts of sunlight striking a tranquil lake shine in our eyes because they are reflecting off the sharp boundary between air and water. Similarly, a sudden change from one type of rock to another inside the planet will cause seismic waves to bend or even reverse their paths.

This unusual property is what first alerted researchers to the fact that a core exists at the center of our earth. The Irish geologist Richard Dixon Oldham discovered the core in 1906, when he noticed something odd about the seismographic records of a large earthquake that had shaken Guatemala four years earlier. What he noticed was that something inside the earth was casting a shadow. The shadow was in the shape of a ring about two-thirds of the way across the planet from the earthquakes' source. Within this ring, deep-traveling P waves seemed to never make it back to the surface. Yet both inside and outside the ring, P waves showed up as usual.

Oldham soon realized that P waves diving into the earth were encountering the boundary between the core and mantle.

P waves either bounced off this boundary back toward the surface or bent toward a point on the opposite side of the globe, thus leaving the distinctive ring-like shadow. And that shadow stood out clearly to Oldham because the boundary that caused it was so sharp. It marked the point where the solid rock of the mantle changed into molten iron.

That metal core formed early in the earth's history, when the world was young and just beginning to heat up. All the iron in the planet turned to liquid and, because of iron's great weight, the tiny droplets sank to the center of the planet. Swirling currents of molten metal collected in a central ball that spanned nearly half of the earth's radius.

There the iron sits today. At the center of the earth, the temperatures reach thousands of degrees Fahrenheit. Yet in spite of this tremendous heat, geophysicists now know that the iron in the core didn't stay completely molten. Once again seismology illuminated another layer of complexity within the earth's interior.

In 1928 and 1931, two large earthquakes in New Zealand produced strong traces on seismographs throughout the world. Examining these was the Danish seismologist Inge Lehmann, who was puzzled about something she saw. A few P waves had arrived at stations that should have been within Oldham's shadow zone. They should have been deflected by the core. But Lehmann believed that these waves were, in fact, traveling part of the way through the core and then bouncing off a sharp boundary somewhere inside it. In 1936, she proposed that the earth had an inner core as well as an outer core. This inner sphere would be solid iron in the midst of the molten metal. Despite the high temperature at the center of the planet, the inner core was being squeezed into hardness by the enormous pressure.

With Lehmann's insight, geophysicists began to see that

the earth was somewhat like a Russian doll; it was made up of many neat, concentric spheres. The planet had a solid inner core and a liquid outer core, both metallic. Together they made up just over half of the planet's 3,960-mile depth. And wrapped around them were the rocky, outer layers of the earth: the thick shell known as the mantle and its thin skin of crust.

Basing their work on seismology, geophysicists quickly divided these outermost regions into layers as well, using not just the strongest earthquakes but quakes of all sizes. The trick was to gather the traces of an earthquake from ever more distant seismic stations. That way researchers could see the waves passing through ever deeper layers of the mantle. From this they could determine the average speeds the waves traveled at different depths. That speed told geophysicists something about the condition of the rock the quake had traveled through. Depending on how dense the rock was or how rigid, for example, seismic waves would speed up or slow down.

Thus seismologists divided the mantle into two main layers, upper and lower. The lower mantle, which extends from a depth of 400 to 1,800 miles, is under extraordinary pressure. Seismic waves travel faster and faster the deeper they dive, reaching a peak speed of some 30,000 miles per hour roughly 150 miles above the core. Just there, where the lower mantle meets the core, the waves encounter a patchy layer that scatters them in ways that have puzzled several generations of geophysicists.

But the upper mantle, which is crisscrossed by both surface and body waves, is far better understood. Seismic waves show the rock to be under much less pressure. In fact, between about 45 and 155 miles down, the rock in the upper mantle is soft—probably partially molten. Geophysicists have dubbed this soft layer the asthenosphere, from the Greek *a,* "without," and *sthenos,* "strength." It lies just below the layer of rigid tectonic plates, which geophysicists call the lithosphere, from the Greek *lithos,* meaning "stone." Together, the lithosphere, asthenosphere and deeper regions of the upper mantle are like the parts of a jelly sandwich. The asthenosphere acts as a kind of cushion between the two stiffer layers, allowing the tectonic plates to slide about the surface of the earth.

What divided the upper and lower mantle was not clear to seismologists. Between 250 and 400 miles down, seismic waves slowed abruptly, marking what geophysicists refer to as the transition zone. This is the much-debated layer that Stephen Haggerty glimpsed when he picked up an ultradeep xenolith from a South African kimberlite pipe.

Together the varied layers are the modern geophysicist's seismic legacy. Without ever seeing the inner depths of the planet, researchers had assembled this model step by step, each generation building on the work of the previous one. Using the rock from kimberlite pipes, they worked out the composition of the mantle and the changes these minerals would undergo at deep-earth pressures, figuring out, for example, that the predominant mineral in the upper mantle, olivine, would convert to an even denser material known as perovskite when it encountered the intense pressures of the lower mantle.

By 1970 the layered model was so successful that researchers could predict to within about a second when an earthquake's tremors would arrive at any seismic station on earth. Yet with the advent of plate tectonics, this model had to evolve. The fact that the earth's surface was moving implied that there was more to its interior than spherical shells. If, as researchers believed, plate motion was tied to a churning mantle, then disturbing these layers were powerful currents of rock.

It was as if scientists had spent decades figuring out the layers of the earth's atmosphere—troposphere, stratosphere, and so forth—and then had suddenly found out about wind.

What were the effects of these subterranean "weather" currents? What high-pressure or low-pressure rock fronts were stirring up the surface of the globe? On top of all this was the problem of hot-spot plumes, which seemed to bypass both the mantle's layers and its roving currents.

All the details of these moving currents of rock lay between the lines. Researchers could still depend on the work they had done, sorting out the earth's layers and their average traits. But within that one second of leeway in their calculations of seismic wave speeds was a wealth of information that geophysicists could exploit.

Each time the speed of seismic waves strays from the average, researchers learn something about the rock those waves have passed through on their journey. Patches of the mantle that are slightly hotter or slightly more fluid will slow the waves down. Similarly, colder, stiffer rock will cause them to speed up.

These fluctuations in the wave speeds are remarkably subtle. But the tiny changes held out the possibility of seeing the churning currents of the mantle. And thanks to two recent advances, researchers had hopes of probing these tiny anomalies. First, the new digital seismometers had unleashed upon geophysicists a flood of sensitive, accessible data. But more important, a handful of ingenious researchers dreamed up a way to stroke those data. Their unique approach paved the way for an extraordinary leap into the deep earth.

Seismic Tomography

One of the architects of this new technique was Harvard geophysicist Adam Dziewonski. Not a particularly casual guy, he seems out of place in the pastel windbreakers he often sports. On the contrary, he comes across as being a bit stiff, so seriously intent is he on his work. But all of Dziewonski's diligence has paid off. The seismic map for which he's known is one of the biggest advances in earth science in the last two decades.

Dziewonski was one of the first to realize that to see into the body of the earth, geophysicists could borrow a technique that physicians had used for years to see into the human body, namely, computerized axial tomography—the CAT scan. In a CAT scan, an X-ray source is rotated around a patient. The X rays travel through the body and out the other side, where they are picked up by a bank of detectors. But some of the rays come to a dead end inside the body's tissues; and some tissues, like bone or tumors, absorb more X rays than others, casting a darker shadow on the detectors. Physicians take all of this information from many intersecting rays, feed it into a computer, and reconstruct a three-dimensional, tomographic map of a patient's internal organs.

Using seismic waves as their invisible probes instead of X rays, geophysicists adopted a similar approach to imaging the body of the earth. In seismic tomography, as the technique was called, researchers focused on the travel times of waves. A seismic wave passing through, say, a warm current in the mantle would take slightly longer than average to complete its journey. By combining the travel times of tens of thousands of crisscrossing wave paths, researchers hoped to map the earth's churning innards as well as a CAT scan maps their own.

Dziewonski published the first map of the mantle based on seismic tomography in 1977. "It was accepted," he says, "with great skepticism." His colleagues were not convinced that he was mapping anything more than noise in the data. Even Dziewonski must have realized that there was room for improvement. Shortly after that first map came out, he seized on "an idea of how to do it better." Dziewonski's second attempt benefited from the wealth of digital data coming on-line. With Harvard colleague John Woodhouse, who's now at Oxford University, he set to work on an improved mantle map.

Across the United States, another team was racing to do the same. That team was headed by geophysicist Robert Clayton, a sage, rumpled professor at the California Institute of Technology. Together with Robert Comer, now at the Analytic Sciences Corporation in Reading, Massachusetts, Clayton began doing his own brand of seismic tomography in the early 1980s, using a technique that was slightly different from Dziewonski's. And so it was that two sets of seismic tomography maps of the mantle arrived in a big splash nearly simultaneously.

"Nineteen eighty-three was the breakthrough year when the mantle became a very hot subject," says Dziewonski, smiling at his own pun. Both the Harvard and Caltech groups took their maps on the lecture circuit, giving talks at scattered conferences and universities. As they toured the country showing the fruits of seismic tomography, they fanned the flames of interest in this new technique. Word of it spread like wildfire among geophysicists.

Seismic tomography showed once and for all that the mantle was indeed in motion. By way of illustration, think back to the psychedelic sixties, to lava lamps with gooey, colored blobs rising up from the heat of the light bulb beneath them and cold blobs falling down toward it. The mantle is not wax and colored water; but fueled by heat from the core and from within, it mixes in much the same way. Hot mantle rock ascends hundreds of miles toward the earth's surface while cold rock dives toward its interior. This steady churning underfoot has not ceased since it began over four billion years ago. Yet in this precious instant of their tenure on earth, researchers could finally see it. They could now state with confidence that the basic theory of convection underlying the roving plates—a theory that had held credence for only a few decades—had at last been borne out.

In the same way that studies of the seafloor overhauled the vision of the planet's crust more than twenty years ago, this glimpse of the power and tumult inside the world heralded a new age in the understanding of the earth. With visionary aplomb Dziewonski and geophysicist Don Anderson at Caltech wrote their predictions in an article for *American Scientist:* "We believe these results signify the beginning of a revolution in earth science. For the first time we can see details in the deep interior of the earth that cast new light on the origin of features observed at the surface, such as mid-ocean ridges and the enigmatic 'hot spots,' like Hawaii and Iceland."

From surface to core, the great engine of the earth is constantly churning to rid itself of heat. Researchers knew that the mother lode of the earth's heat came from the radioactive decay of mantle rocks and the heat left in these rocks since the earth's formation. How this heat dissipates, how it causes the rock to churn, is the blood and guts of the animate earth. It is the key to all of the large-scale changes on our planet. Now it was laid out before them in the great Technicolor blobs of the tomographic maps.

Smooth blobs as orange as flames marked the spots where seismic waves passing through the mantle slowed down and thus where geophysicists assumed the rock was hotter than average and rising. Similarly, blue blobs signified cold rock sinking toward the interior. Hanging on the walls of Dziewonski's office at Harvard are framed images of these original tomography maps, which show the temperature profiles of different layers. As we spoke one sunny, winter day, he proudly motioned to them, pointing out the key details.

His hand moved round and round, tracing the outline of a blue ring. It is one of the most striking features of Dziewonski's work, an unmistakable circle of cool rock that is apparent in maps from the surface to just above the core-mantle boundary. It underlies the borders of the Pacific Ocean, beneath the famous Ring of Fire. Here slabs of crust dive beneath the continents, stirring up a wreath of volcanoes on shore. But tomography picks up only a chill echo of the fires at the surface—a cylinder of cold, sinking crust.

Dziewonski's maps also showed fat, warm blobs perched beneath the Mid-Atlantic Ridge, the East Pacific Rise and the Indian Ocean Ridges. Such stores of hot rock were aligned so closely with the magma-belching mid-ocean ridges that they must be fueling them, most researchers agreed. Somehow these shallow reservoirs must provide the stuff that makes new ocean crust.

Yet Dziewonski didn't see these hot patches extending deep into the mantle the way the cold downwellings did. Instead, his maps showed two huge bodies of hot rock dominating the upwellings in the vast lower mantle. These rising currents were like great inverted drips beneath the African Plate and the southern Pacific Ocean. They stood well away from the mid-ocean ridges and bore no relation to any of the boundaries of the tectonic plates or to the pattern of continents and oceans. Seeing no apparent connection to these features on the surface, researchers could no longer envision the upwellings and downwellings in the mantle flowing along some simple loop like the convection cells in the familiar pot of soup simmering on a stove.

An additional snag to their progress in understanding how the mantle mixed was the differences between Dziewonski's maps and Clayton's. The Caltech maps did show a cold ring around the Pacific. But there was little agreement on most of the smaller features. In fact, because of the slightly different technique that Clayton used, his maps looked a good deal messier. He had separated the wave speeds into many small blocks instead of the smooth blobs of Dziewonski's maps, which made it much harder to pick out the large-scale features. Nevertheless, when Clayton's maps were smoothed like Dziewonski's, the differences remained.

Instead of rushing to adopt the results of one of the two teams of tomographers, the geophysics community waited for them to come to some agreement. As it happened, the resolution of this issue never really reached the stage of the researchers ironing out their differences. What started out as healthy competition turned into a victory by default.

Dziewonski and Woodhouse published their maps in 1984. But while the Caltech maps were always available to anyone who asked, Clayton never got around to publishing them in a journal, where all scientific advances must appear in order to be scrutinized by the community. Years later, I went to Caltech to talk with Clayton. Seated in his office, his desk piled high with papers, he quickly ushered me into a clean conference room, murmuring something about getting away from the mess. When I eventually asked him why he hadn't published his maps, Clayton replied honestly, "Well mainly I'm not sure they're right."

Clayton, by his nature, tends to dwell more than most on the uncertainties of his science. As his Caltech colleague Don Anderson once put it, "There's nothing wrong with the data. It's just a matter of sitting down and writing the paper." But Clayton moved on to other projects, and partly as a result, his maps gradually dropped out of the collective debate.

In the years that followed, Dziewonski's maps became the

standard against which geophysicists measured their belief in the technique. Many researchers tried their own hands at tomography and compared their results with Dziewonski's. They examined S waves instead of P waves, surface waves instead of body waves. What they found and what has gradually convinced them of Dziewonski's genius was that the key features in his original maps kept reappearing: the cold circle around the Pacific, the two hot upwellings in the lower mantle.

As more data were accumulating, Dziewonski's maps were going through successive upgrades. With each new model the resolution sharpened and the individual rock currents became more distinct. "As our confidence that we had seen real things grew, we invested some time and effort in graphics," said Dziewonski at a recent retrospective on studies of the earth's interior. His tone was carefully laced with humility, a symptom of the difficulty so many scientists have presenting "flashy" results to their colleagues. Indeed, Dziewonski's latest version of the maps are visually arresting images that resemble Henry Moore sculptures. Three-dimensional blobs slip down into the mantle or rise like fat flames from the surface of the core.

Lately Dziewonski has begun giving names to the grand structures of the earth's interior: the China High, the Great African Plume. His names conjure up the days when explorers such as the Scottish missionary David Livingstone scoured the Dark Continent in search of the source of the Nile. To Dziewonski, as to geophysicists in general, the features in his maps are the modern landmarks of the interior as surely as the great rivers and mountain ranges are landmarks of the surface.

Those features have been imprinted on researchers' minds to such an extent that in 1991 Dziewonski began a paper written with Harvard colleague Robert Woodward invoking the claim made in his earlier *American Scientist* article. There he had proposed that seismic tomography signified a revolution in earth science. "We can ask now, seven years later, whether this promise has been kept," Dziewonski and Woodward continued. "The answer, with some correction for subjectivity, is 'yes.' "

Most geophysicists would say there was not just one advance or one group of researchers that brought about the revolution in the way they think about the world. Instead, many separate insights and discoveries took place within the decade of the 1980s. These pieces fit together in unforeseen ways, presenting researchers with a radically new vision of their planet's interior.

Yet throughout it all, the tomographic maps provided the foundation on which these other discoveries were raised. Seeing into the mantle had forever changed the way that geophysicists thought about the great mass of the planet's interior. Thus with tomography began the modern journey into the earth. In the years to come, researchers would find in those colored blobs connections to a host of structures even deeper inside the world.

4

ANTI-MATTERS

Drive west across the northern Great Plains states and you're entering Big Sky country: Montana, Wyoming. The cloud cover is so high that it feels as if someone has reached down and lifted the lid off the sky. Drive farther west, past the Rocky and Cascade mountain ranges into coastal Oregon or Washington, and the weather turns to a constant haze. Enough rain accumulates in a year to drown a standing man.

The reasons for Seattle's cloudy skies and Cheyenne's clear ones are the mountain chains that lie between them. Heavy, turgid air from the west meets the peaks of the Rockies and the Cascades, and before crossing into the northern plains it dumps its load of rain on the coastal states.

As this stream of air makes its way around the globe, it is stalled by dozens of similar hurdles: the Atlas Mountains of northwest Africa, China's Tien Shan range and the Great Dividing Range of Australia, to name a few. On either side of each barrier it leaves similar pockets of high and low cloud cover, of arid and lush terrain. But in addition to its effects on the world's climate, the wind pounding against the mountain ranges also has an effect on the world.

The breezes are like a puff of air on a pinwheel. They catch on the earth's mountain ranges and temporarily cause the planet to spin a bit faster. As a result, our days sometimes go by a shade quicker than twenty-four hours. In records kept by the International Earth Rotation Service in Paris, it's apparent that from season to season the days grow shorter and then longer again by about a thousandth of a second, in a regular sawtooth pattern.

"In that same data there's a more curious function, to me at least," says Caltech's Robert Clayton, as he points to the broader undulations in the jagged line. "There's a cycle moving around on about a decade. For example, at the moment the earth is currently slowing down. I'm talking something on the order of four milliseconds a day. In five years it will temporarily speed up."

"What's causing that cycle?" he asks rhetorically. "It cannot be the atmosphere." Weather patterns don't come and go on a decade-long timescale. Clayton and other geophysicists believe that the cause lies deep in the earth, where a different kind of wind is pushing against a different kind of mountain.

Eighteen hundred miles below the surface, at the base of the mantle, lies one of the earth's most extraordinary terrains. Contrary to the idea of the earth's layers set neatly on top of one another, geophysicists envision a rugged landscape in its depths. They speak of anti-mountains, clinging to the underside of the mantle the way stalactites hang from the ceiling of a cave, and anti-valleys standing between them. Some researchers believe the topography may even approach a Himalayan scale, spanning several miles from peak to trough. And flowing almost as quickly as air across this terrain are the swirling currents of molten iron in the core. These iron winds push against the anti-mountains at the core-mantle boundary. They are the cause of the decade-long changes in the length of our days.

It is almost as if the Roman philosopher Seneca was right, that "there exists below everything that you see above." What exists at the intersection of the core and mantle may be a strange, inverted landscape that in many ways resembles the landscape at the surface of the earth. It is not a meeting of land and sky, but it is a meeting all the same. This deep terrain is subject to all the same processes of weathering and chemical interaction that take place at the surface. Everything from erosion to leaching to iron rainstorms is possible, perhaps much more.

Because of the extreme conditions in the earth's interior, the range of interactions is much broader than at the surface. The sheer contrast between the layers of core and mantle is enormous. The jump in density across the boundary is twice what it is where the solid earth meets air. Add to that the exaggerated heat and stresses at that depth, and you're entering the realm of the unimaginable. "It's at inconceivably high temperature and pressure," says University of California geophysicist Elise Knittle, who studies this boundary layer, "way over a million atmospheres and temperatures between four and five thousand degrees kelvin [6,700–8,500 degrees Fahrenheit]."

This elaborate anti-landscape came to geophysicists' attention just around the time that seismic tomography was showing them the churning mantle, and added a dose of intrigue to studies of the earth's interior. Like Alice of the children's tale, geophysicists happened upon a Wonderland. Curiouser and curiouser it grew the closer they looked.

In some cases, researchers had only recently acquired the

tools and techniques to probe this deep boundary thoroughly. Other revelations came from taking a fresh look at the data. These were just the sort of keys that geophysicists needed. As the door to understanding this Wonderland opened a crack, many researchers rushed in. Seismologist Thorne Lay remembers the "surge of enthusiasm and redirection of people's efforts. The core-mantle boundary topography captured a lot of people's imaginations."

Lay himself deserves much of the credit for awakening geophysicists to this wondrous terrain halfway to the center of the earth. Several years ago he turned deep-earth science on its ear with his discovery of distinct structures at the core-mantle boundary, structures that strangely resemble those at the surface of the earth. Lay has since been honored with the prestigious Macelwane Medal, bestowed upon the most accomplished young researchers in earth science, and despite his youth he has already achieved the distinction of having a feature of the core-mantle boundary named after him.

Researchers have known since early in the century that something was odd about this boundary region of the planet, because of the way that seismic waves seem to scatter when they pass through that area. The core-mantle boundary seemed to belong to neither the core nor the mantle. As early as 1949, the bottom 150 miles of the mantle was pronounced a separate layer, known to geophysicists as D″, or D double prime.

Attempts to gain a clearer picture of this boundary layer floundered for several decades. Some studies showed the seismic waves that passed through it speeding up; others showed them slowing down. But during the 1970s and early 1980s, geophysicists gradually moved beyond their mode of thinking of the earth in terms of layers. As they started thinking about the variations within those layers, research on the core-mantle boundary grew. This was particularly true in Europe. German and Dutch researchers had been paying much more attention to this boundary region than geophysicists in the United States when Lay began his Ph.D. thesis on it.

Lay used seismic waves to look in detail at patches of the core-mantle boundary. He found places where the waves bounced off the boundary and then returned to the surface near a large bank of seismographs. This allowed him to study many waves traveling along more or less the same path. Once he had increased his resolution in this way, Lay was able to detect a sharp change in the speed of seismic waves at some of the patches on the boundary—a change that has since been called the Lay discontinuity in his honor.

Beneath Alaska, India, Siberia and the Caribbean, seismic waves seem to run into a sharp structure, as if a layer of rock is floating at the boundary. Yet at other sites, like beneath the central Pacific, seismologists have found no such slabs. What has become increasingly apparent over the years since Lay's initial discovery is that the way this unusual boundary layer looks depends a great deal on where researchers look.

"It's very analogous to what we see at the surface of the earth," says Lay. Large-scale variations exist between oceans and continents, and seismologists see these differences reflected in their seismograms. What Lay discovered in the deep mantle were "coherent, large-scale features—pretty much like continents." Anti-continents, news stories began calling them. One such slab under Alaska is 170 miles thick and some 950 miles across, or roughly the size of Saudi Arabia.

These anti-continents provided an arresting image of the earth's interior, and one that came on the heels of the vivid maps of seismic tomography. Lay believes the joint timing of these two breakthroughs drew a great deal more interest toward his boundary-layer research than might have come otherwise. Yet interest is by no means consensus.

"Some of the most vocal critics were the Germans, who were working on this continuously throughout the sixties and seventies and into the eighties," says Lay. This was their turf, and they were upset with an upstart claiming to have seen entire continent-sized masses they had missed. Lay calculates that he spent three years just trying to respond to criticism, to show that alternative explanations weren't valid. He felt all along, though, that he was on solid footing. The data had been around for twenty years. He had just discovered a new way to plot them.

Lay's fresh approach to the seismic data gave a tangible shape to previously vague ideas about how the core-mantle boundary layer looks. As a result of his efforts, a majority of geophysicists are now believers. Not all of them have lined up behind Lay's idea of anti-continents, but most researchers would argue that there is some sort of terrain down there, begging for further scrutiny. And that is what they are concentrating on today. Since Lay's results first came out, the seismological tools for seeing the core-mantle boundary have grown more sophisticated. With each passing year, geophysicists get closer to mapping the layer in fine detail.

In the last few years, for example, seismologists John Vidale and Harley Benz at the U.S. Geological Survey in Menlo Park, California, have carved out a promising new niche in seismology. California is riddled with seismometers that geophysicists have set up in recent years to take the pulse of the San Andreas Fault as well as other faults in that region. Similar banks of instruments are set up throughout the United States and Canada for local monitoring. But Vidale and Benz are using this array of instruments like a broad net to collect seismic data and get a better look at the deep interior.

On May 21, 1992, Chinese researchers conducted an underground nuclear test near Lop Nor, a lake in the northwest part of the country. The sharp pulse of energy traveled down through the earth just as earthquake waves do, and those waves that grazed the core-mantle boundary beneath Alaska and the Arctic Circle were charted by Vidale and Benz's array. It was a rare opportunity, a magnitude 6.5 pulse, strong and precisely timed. It provided the best resolution yet of that boundary layer. Under the northwest tip of Alaska, Vidale and Benz were able to make out a slab of anomalous, probably cold, material. At 185 miles across and 80 miles thick, this patch of rock hovering near the boundary was more like an anti-island than an anti-continent. With the seismic array he was using, Vidale says he could tell that very close by there was no structure at all.

Vidale believes that such detailed study of the core-mantle boundary shows the motley nature, either in the chemistry or the other physical properties, of the rock deep within. Part of understanding how this boundary layer looks involves understanding what it is made of. With that knowledge researchers could approach even bigger questions, like the origin of these anti-continents. Yet the makeup of this jumbled layer at the base of the mantle continues to elude geophysicists.

The Boundary in Miniature

Some researchers believe that the anti-continents are not only a parallel to the surface but might actually be made of the stuff *from* the surface—that is, sinking slabs of oceanic crust. Lay has compared the positions of these anti-continents with the tomographic images of the mantle and says that they tend to be beneath regions of the mid-mantle that are cold and sinking. Some researchers take this as a sign that ancient pieces of oceanic crust may be falling to the core boundary and settling there, forming the anti-continents.

Less dramatic is the idea that the anti-continents are simply mantle dregs—heavier materials that came to rest above the much denser core in the way that particles settle between the oil and vinegar in a salad dressing. These dregs might easily form a patchy, uneven layer. They might be billowing about in the mantle's currents like snowdrifts in a storm, and each pile of dregs could show up on seismic maps as a single landmass.

But perhaps the most startling discovery was that, instead of falling to the boundary, the patchy layer may simply grow out of it the way rust grows atop cars in a junkyard. Such chemical reactions take place continuously at the earth's surface. Minerals in the rock react with the atmosphere and the environment to alter the landscape. In fact, to stretch the analogy further, there is a sort of rusting that goes on in the soil—iron interacts with oxygen in the atmosphere to produce the reddish tint we associate with clay. A similar chemical reaction, researchers discovered, could be taking place between the mantle and the core.

That realization stems from a groundbreaking experiment by Santa Cruz geophysicist Elise Knittle. One day while talking to Knittle, I asked her to give me a tour of her lab and show me the apparatus she uses for her deep-earth studies. She led me down a hallway and, with a chuckle, pointed at a lab bench. "This is it," she said, reckoning on my disappointment. Perched on the bench was the tiny instrument known as a diamond anvil.

Still the hardest substance known to man, diamonds enable researchers to squeeze samples of earth materials to the pressures they endure deep in the earth. Picture two gem-quality diamonds, each tapering to a point. No wider than pins, these highly polished points are placed end to end, and a mineral sample about a third of that size is sandwiched between them, whereupon the diamonds are used like the opposing arms of a vise.

Knittle picked up the shiny anvil and rested it on the palm of her hand. The beauty of this instrument is that by conducting experiments on a miniature scale, geophysicists can bring enormous force to bear on the surface area of the minerals. The entire device is held together by a metal frame roughly the size of a silver dollar. Simply by hand-turning a few screws around its rim, geophysicists can gradually tighten the squeeze.

The diamonds, in addition to being tough enough to endure these pressures, are also conveniently transparent. This means that researchers can view a sample as it is being compressed. They can watch the rock change color or become shinier under pressure. Most important of all, they can heat the sample through the diamond window using a laser beam, achieving the high temperatures that accompany high pressures in the earth.

Knittle set up this diamond anvil lab a few years ago with her husband, geophysicist Quentin Williams. They modeled it after the one where a landmark core-mantle boundary discovery was made: Raymond Jeanloz's laboratory at the University of California in Berkeley. There Knittle, Williams and Jeanloz had set out looking for the answer to a completely different question, another example of how often science progresses by serendipity.

In 1986, when Knittle was a graduate student at Berkeley, Williams and Jeanloz were conducting a series of experiments to measure the melting temperature of iron at pressures approaching those in the earth's core. Knittle was working on a related project. Researchers believe that the core is not pure iron, but is mixed with various elements. So Knittle was mixing iron with different minerals in the diamond anvil to figure out how these impurities would affect the metal's melting temperature.

She punched tiny discs from a flimsy sheet of iron foil and placed them in the anvil. As part of the experiment, she packed the iron foil in a fine dust made of ground rubies. Minerologists have long known that iron and ruby will not chemically react with each other at moderate pressures. That's why Knittle was using the ruby dust as an insulator. But at the high pressures of the experiment she noticed something unusual: the iron *did* react with the ruby. Given enough pressure and high-enough temperatures to melt the iron, "it reacted so extensively with the ruby that essentially you didn't have any ruby there anymore," she noted. In its place was a halo of entirely new material.

Chemically, ruby is similar to perovskite, the material thought to make up the lower mantle. To Knittle, the connection with the deep earth was immediately clear. These reaction halos, she said, were "almost a core-mantle boundary in miniature." Packing the iron foil in perovskite, Knittle repeated the experiment, and again it "reacted like crazy" once the pressure was high enough.

Knittle and Jeanloz proposed that what happened on this miniature scale could be happening extensively deep inside the earth. Reaction products could be piling up at the core-mantle boundary. Like the rust on a junkyard car, that boundary layer could get only so thick. Somehow, the reaction products would have to be continuously scraped away and pristine mantle exposed to the core in order to build up a layer of any depth. But that scenario presented geophysicists with a radical yet plausible explanation for the anti-continents.

It was a remarkable discovery, a previously unknown

chemical process that Knittle had stumbled onto. Experiments like this opened geophysicists' eyes to the myriad ways this boundary layer could look. Yet the problem researchers have faced in the years since Knittle's discovery is an inability to narrow down these choices.

Strapped by the limitations of modern techniques, they still have numerous questions about how the core-mantle boundary layer looks, how it got there and how it behaves. Is it a patchy layer of aggregate rock, either mantle dregs or reaction products, or some combination of both? Or does it cover the core like a layer of icing, thicker in some places and thinner in others? Do the gales of mantle convection sweep the layer into piles, or is the boundary layer itself convecting, generating standing waves?

Every deep-earth geophysicist has his or her own version of this subterranean landscape. They pour their ideas into crude sketches—"cartoons" they call them—that show the earth in clean cross-section. Big blobs and little blobs float at the core-mantle boundary, slabs of material fall toward it or hot jets rise off it. Sometimes questions marks float next to parts of the picture where researchers aren't quite sure what to draw.

Compared to the slick colored maps of seismic tomography, these cartoons seem like country cousins. But in terms of the reasoning that goes into them, they are state of the art. They are the charts of a new frontier, modern-day versions of the maps made before ships circumnavigated the globe. Like the ancient mapmakers, today's geophysicists are describing the routes that future explorers will follow.

5

SPIGOTS OF FIRE

Many geophysicists envision the core-mantle boundary as a sort of deep-earth launchpad. They believe it is the source of hot-spot plumes, the mysterious jets of magma that J. Tuzo Wilson imagined rising through the mantle, poking holes into the moving sheets of crust. Hawaii and all the other hot spots on earth have their roots in these unusual spigots of fire that siphon intense heat off the core and carry it to the surface. Now geophysicists believe that plumes may be having a greater impact on the earth's surface than Wilson ever realized.

The vision of plumes persists in spite of the fact that plumes are almost as much of a mystery today as when they were first conceived. In all the years since Wilson proposed his theory of hot spots, plumes have remained elusive, practically imaginary structures. No one has ever imaged a hot-spot plume with seismic tomography. The best that researchers have done in recent years is to detect what looks like the head of a plume rising off the core-mantle boundary.

Eighteen hundred miles beneath Indonesia, Washington University geophysicist Michael Wysession and his colleagues recently found a particularly slow patch in the paths of seismic waves that are known to bounce along the core-mantle boundary. Presumably, the waves were slowing down as they passed through hot rock, and Wysession says the hot patch appears to be slightly rounded, as if a plume is just beginning to bud off the boundary layer. "What I've seen underneath Indonesia is the closest thing that anyone has seen to what this would actually look like," Wysession says. He has yet to prove, though, that this patch is the real McCoy, the birthplace of a hot-spot plume.

Other researchers have noticed what they think is a plume beneath the Bowie Hot Spot, a submerged volcano off the west coast of Canada. Seismic waves from Alaskan earthquakes cross under this spot before reaching detectors in Washington State. At 435 miles below the ocean floor, the waves show a small but distinct slowdown, a warm patch. Oddly enough, this ninety-mile-wide patch lies about ninety miles to the east of the hot spot on the surface. That doesn't rule it out from being part of the plume beneath this hot spot, but as evidence it is far from conclusive.

Proof, most researchers agree, would amount to a picture of the object fully formed. A plume caught in the act of burning its way up through the mantle would convince any doubters that these structures exist. Yet hot-spot plumes are like slippery eels refusing to be caught; they are intrinsically the hardest things for seismologists to detect. That is partly because the structures are small—compared with the six-hundred-mile-wide blobs in seismic-tomography maps, hot-spot plumes are probably only a couple hundred miles across. In addition, plumes have remained undetected because they are hot.

Since seismic waves travel more slowly through hot material, often the quickest pathway for a seismic wave to take is to just go around the plume, much the way a driver gets home quicker by going around a traffic jam rather than pushing through it. If the seismic waves miss the plume, their record of it is lost to geophysicists. This explains why so many researchers who have gone searching for plumes have come up empty-handed.

Nonetheless, geophysicists continue to look, continue to put their faith in ghosts of a sort. What convinces researchers that these things exist, and moreover that they know how plumes should look even though they've never seen one, is indirect evidence. You might say they don't need to see the skunk behind the tree to know it is there. Sometimes pieces of evidence can mesh together so well that they begin to resemble proof, and the scientific community becomes hard-pressed to explain the body of evidence any other way.

In the case of plumes, the evidence for how they look comes from humble origins. In a series of crude but pioneering experiments, geophysicist John Whitehead raided the kitchen cabinet, so to speak, finding substitutes for the stuff of the earth in common household objects and, in the process, laying the foundation for modern ideas about the roots of hot-spot volcanoes.

Whitehead, known as Jack to his friends, has a solid frame. His sandy hair is short, almost a crew cut, and the way the skin crinkles around his eyes gives away the many years he has spent in salt air. He works at the Woods Hole Oceanographic Institution on Cape Cod, Massachusettes, a sprawling collection of research labs and offices that all lie a stone's throw from the clanging halyards and squawking seagulls of the Woods Hole harbors.

Whitehead's own laboratory is filled with what at first

glance resemble glass fish tanks. But they are actually so narrow that they wouldn't leave a fish room to turn around. Instead he fills them with any of the dozens of different varieties of liquid that line the walls of his lab. From Karo syrup and glycerin to common motor oil, these liquids are the tools Whitehead uses to mimic the flow of rocks inside the earth. "I amuse myself by thinking these lab experiments are a little bit like a time machine," he says, "because we essentially compress time so we can see in the laboratory what takes tens of millions of years in nature."

Many years ago, Whitehead attempted an experiment that even he considered "very off the shelf." While attending an afternoon lecture, he and a colleague got an idea. "We literally went to the laboratory and grabbed a couple of things off the shelf and put them in a beaker and turned them upside down the next morning," Whitehead recalls. In the bottom of the beaker they put glycerin. On top of that they poured a thin layer of silicone oil. They left the fluids overnight to get rid of the air bubbles trapped inside. The next day, when the researchers flipped the beaker, the silicone oil began to form fat mounds.

These protrusions swelled until they eventually budded off the oil layer. They pushed up through the denser glycerin. But behind each of them trailed a skinny conduit, an umbilical cord connecting them back to the silicone oil layer, giving Whitehead's laboratory plumes the odd look of a balloon on a string.

He hadn't been thinking about hot spots at the onset of the experiment. He had thought to replicate salt domes and batholiths, the geological structures that dome up under mountains to create granite. But seeing those little conduits started him thinking about plumes. They, too, were low-density matter trying to buoy up through a denser material.

Using the narrow fish tanks, Whitehead and physical oceanography student Douglas Luther, now at the University of Hawaii, experimented with different fluids of different viscosities in an effort to find liquids that could mimic the dense, stiff rock of the bulk of the mantle and the lighter, less dense fluid of the plumes. Again and again the laboratory plumes would rise up through the tank with that familiar shape of a balloon on a string. The researchers then spent another couple of years working out the mathematical equations that these odd structures should obey in the earth's interior. But by the time they published their findings, in 1975, interest in plumes had already begun to languish. "Nobody really thought about them much," Whitehead says. "I finally got discouraged." Almost a decade would pass before new progress was made.

"When I got interested in plumes, about six or seven years ago, it was a subject that was held in disrepute," says Peter Olson, a lanky, moon-faced geophysicist who at that time was just starting his career at Johns Hopkins University. "There was a group of geodynamicists who either ignored the concept or were actively skeptical about it—that there could be these structures in the mantle. People just didn't believe that was a plausible concept." How could plumes stay fixed and yet originate halfway from the center of the earth?

To complicate matters was the fact that plumes had been such a fashionable notion. Wilson's theory of hot-spot plumes had stirred such interest that researchers began using plumes as a kind of magic bullet to do away with any of a whole host of unanswered questions about the earth. Geophysicists spun theories that implicated plumes in everything from plate motions to where mineral veins arose, with little in the way of proof. Their reckless approach brought on the backlash that occurred in the late 1970s and early 1980s.

But during that time a handful of researchers like Olson continued to believe that much work still needed to be done to understand these bizarre structures. As Olson recalls, a few geophysicists who set out to disprove hot spots ended up switching camps, convincing themselves that these plumes could indeed exist. He says, "What happened is so typical of the way earth-science ideas develop. You can't point to a single critical observation that made everybody change their minds. It was just the accumulation of the evidence. More and more things could be explained by this idea."

As hot-spot plumes once again became respectable, Whitehead's was the model that geophysicists adopted. Even though he had published when interest was waning, he had had dozens of requests for copies of the article—a great many for a scientific paper. A decade later, his crude tank experiments laid the basis for new work.

Researchers realized that hot-spot plumes wouldn't actually be liquid flowing through liquid. In fact, the plume rock would only melt near the surface. Though the concept is difficult to envision, the plume would be solid rock flowing through solid rock most of the way up through the earth. Yet even if plumes were solid, the crudeness and simplicity of Whitehead's experiment only added to its power. Scientists could repeat the tank simulation and produce the same results time and again. The experiment was "robust," as geophysicists like to say—as hardy as a prizewinning chrysanthemum. Even modern researchers continue to make plumes in the lab, though Karo syrup and blue food coloring are the fluids of choice nowadays. The basic results from these updated experiments, as well as more advanced computer models of plumes, are all similar. They reinforce Whitehead's original balloon-on-a-string image of hot-spot plumes.

Footprints on the Crust

Moreover, that picture seemed to fit with the footprints that researchers saw hot-spot plumes leaving upon the earth's surface. As geophysicists envisioned it, chains of islands like Hawaii would draw their magma from the thin string of the plume. That narrow pipeline could funnel hot rock from the deep earth to the surface for millions of years, burning a trail of volcanoes into the plate moving over it.

But the mother of all eruptions would be the plume's initial salvo. When the contents of that balloon head first burst out onto the surface, it would let loose an enormous flood of molten rock. Like a great spurt of ink, this lava deluge would blot out the oceanic or continental landscape. Only gradually, as the hot spot matured, would this outpouring subside to the dribs and drabs from the plume's string, forming volcanic islands.

The magma flood that inaugurated the most famous hot-spot trace, the Hawaiian Islands, disappeared long ago, having been swept down into the ocean trench by the steady march of the Pacific Plate. Yet in spite of this prominent exception, the history of many other erupting fonts is evident across the face of the globe. With the renewed interest in plumes in the mid-1980s, a virtual cottage industry sprang up trying to link modern hot spots to their tumultuous origins. From the volcanic spouts that mark the volcanoes' positions today to the extraordinary magma floods that announced their arrival at the surface, geophysicists found they could catch plumes spouting throughout the eons.

Over 41,000 cubic miles of lava flooded the surface of southern Washington and northern Oregon 17 million years ago to produce what is known as the Columbia Plateau. That hot-spot eruption lasted some 2 million years before petering out to a more modest flow. Now the hot spot lies beneath Yellowstone National Park, fueling the geysers, mud pots and volcanic domes that draw thousands of tourists to the area.

An even more fantastic eruption took place 65 million years ago in western India, heralding the arrival of the hot-spot plume that lies beneath Reunion Island today. In only half a million years, some 350,000 cubic miles of lava flooded the continent. Great pancakes of lava piled up layer upon layer, rising over a mile and a half in some places. The same eruption in North America would have covered California with a layer of rock two and a half miles deep. In India it left behind one of the largest volcanic deposits on earth, a series of stepped lava terraces known as the Deccan Traps.

Many other examples of hot-spot floods exist around the world. Such wholesale repaving is certainly damage enough. But geophysicists who supported Whitehead's balloon-on-a-string model also realized that a plume's arrival at the surface could have even more dramatic repercussions.

Picture the rising head of a hot-spot plume flattening like a pancake as it reaches the earth's crust. The ground above it swells from the heat like a shirt straining against the belly of a fat man. And eventually, those shirt buttons are going to pop. For similar reasons, deep rips in the crust often accompany the arrival of plumes at the surface. One of the most potent examples in our modern era is taking place right now in eastern Africa.

The African Plate looks as if it has drifted over an open fire. It is pockmarked with over forty active or recently active hot-spot volcanoes, roughly a third of the world's total. Camped above one of the largest reservoirs of hot mantle rock, this continent's crust is swelling from the heat. But having so little elasticity, that crust has also begun to crack.

In eastern Africa today, two long, narrow valleys connect a series of finger-like lakes. One valley runs north from Lake Nyasa in Malawi through Tanzania and Kenya to its end in Djibouti. The other curves around the western side of Lake Victoria, through Lakes Albert, Edward and Tanganyika. These valleys are known as the East African Rifts. They mark the joints where huge blocks of the African continent are starting to break away from the mainland. Before our eyes the land is literally being ripped apart.

As the rifts widen, their side effects spring up across the countryside. The boundaries of the lakes are continuously evolving, shifting water supplies and fishing sites in the dozen nations they cross. The volcanoes that line these tears in the crust are coating the land with lava, building up layer upon layer all along the rift valleys. And in the midst of these rift zones lies the Serengeti Plain, a land of plenty for African safari hunters.

Enormous herds of wildebeests, zebras and gazelles streak across the savannah, cheek to jowl with packs of their predators. They have all been trapped by the high rift walls that fence in the plain. Their numbers testify to the impact of the breaks, not only on the land but on the wildlife that inhabits it.

Yet rift impacts have been plaguing this part of the globe since long before man, or even animals, walked upon it. Some 20 million years ago, a bout of continental rifting opened up the Red Sea. In the beginning, that sea looked just like the East African Rifts—a long line of unconnected lakes. Now it is an afternoon's crossing by boat.

About 10 million years before that, another rift began along what is now the southern shore of the Arabian peninsula. Today the Gulf of Aden is a full-fledged ocean by tectonic standards. Beneath its waters, magma from the mantle is escaping through cracks to the seabed, continuously soldering new crust onto the sides of the break. But fields of lava throughout Yemen, Ethiopia and western Saudi Arabia testify to this region's initial rifting. All told, the cradle of civilization has been tearing asunder for some 30 million years.

These great rifts, together with the magma floods that often accompany a plume's arrival at the surface, gave researchers some confidence that their balloon-on-a-string image of plumes was accurate. That picture began to solidify in their minds just about the time that geophysicists were also grappling with the discoveries of the mid-1980s. The maps of tomography, anti-continents at the core-mantle boundary and hot-spot plumes presented a monumental profusion of ideas.

Geophysicists quickly discovered parallels between the other structures of the deep interior and plumes, which spanned the earth's layers from crust to core. More links between the distant parts of the globe began to sprout up all over. Geophysicists could see how the activities of the deep impinged upon the surface. And as a result, they found themselves looking at the earth from a new perspective.

The Gatekeeper

Think back to geophysicist Adam Dziewonski standing next to the tomographic map of the mantle hanging on his office wall. He waves his hand round and round to show the globe-encircling blue ring of cold material sinking into the earth. On either side of this ring are two warm upwellings, the great inverted drips of the lower mantle. Overall, these upwellings and downwellings form a pattern on the earth much like the striped pattern on the surface of a billiard ball.

That same pattern turned up again when geophysicists looked at hot spots. Nearly all the hot spots on earth fall within two distinct clusters. One of those clusters lies beneath Africa. The other is across the globe in the central Pacific, where not only the familiar Hawaiian hot spot sits but also a group of less renowned plumes, including those that created the Pitcairn Island group and the Carolines. Soon after the seismic-tomography maps came out, researchers discovered that the two main upwellings in the deep mantle matched up with these two clusters of hot-spot plumes.

A curious coincidence? That was one possibility. But most researchers were inclined to see a connection between the narrow plumes and the broad, warm patches of the lower mantle. Naturally, this raised new questions about how and why these structures interacted in the deep earth. Perhaps the balloon plumes could rise only in these warm patches in the mantle, or perhaps the plumes were just fooling tomographers into thinking the mantle around them was warm. Either way, the truly important insight for geophysicists lay in seeing that rock currents deep in the mantle were having a tangible impact on the earth's surface. It turned out that that impact was not limited to volcanic eruptions.

In the mid-1980s, Massachusetts Institute of Technology geophysicist Brad Hager and his colleagues showed that the hot-spot clusters and warm mantle upwellings also matched up with another feature: the pattern of bulges in the not-quite-spherical globe. Two circular bulges stand on either side of an earth-encircling stripe where the surface is depressed, giving the planet what the researchers called a "billiard-ball–like pattern" of dips and swells. This pattern is readily apparent whenever researchers look at the sculpted surface of the open sea. Although everyone thinks of ocean water as choppy, our mental image of it is nowhere near as rough as it truly is.

Across the globe, sea level varies by hundreds of feet. And these giant undulations are due to more than just waves. Water is attracted to regions where the earth's gravity is higher than average. The ocean literally piles up in those areas where the thickness or density of the earth's rock is concentrated. When geophysicists began measuring the height of the sea surface with satellites in the 1970s, these piles of water stood out in stark relief. This distorted sphere came to be known to scientists as the geoid.

But since gravity erects these bulges in the geoid, it was always a puzzle why they didn't correspond to surface features. The thickness of the crust, for example, varies widely between continents and oceans. It seemed to geophysicists that the continents should shape the gravity field in some way. But after seeing the tomographic maps, researchers began to understand why the continents had no effect. The source of the gravity variations lies deeper. The rising hot rock of the lower mantle pushes up the earth's surface, adding extra mass to those regions. In the process it strengthens the pull of gravity there.

That meant the movements of rock in the lower mantle were literally controlling the shape of the earth. The clusters of hot spots were yet another case of inside affecting outside. And they coincided with the bulges in the geoid, which sent a clear message to geophysicists. The interactions between the deepest mantle and the surface were all somehow tied together by this billiard-ball pattern.

It was a clue to how the planet works, albeit a confusing one. What was creating this pattern? Geophysicists began to think that since hot-spot plumes presumably originated at the core-mantle boundary, this layer might play a role. And thus they scoured the activity of this boundary layer for some insight into the bigger picture of how the layers of the earth work together in the global heat engine.

They began to formulate a role for the boundary layer as a gatekeeper that controls the heat that escapes from the earth's core. Thorne Lay's hints about anti-continents at the boundary as well as Elise Knittle's evidence for chemical reactions told researchers that the boundary layer was probably "patchy." This patchiness might allow heat to flow between the layers, but to flow faster through some areas than others—which might explain how a particular terrain at the core could produce a billiard-ball pattern throughout the rest of the planet.

In plumes, researchers saw structures that might continuously be shaping and reshaping the boundary layer. Each rising plume could scoop up part of the anti-terrain, sculpting it as if it were a lump of clay. There seemed to be no limit to the ways such plumes might be affecting the anti-world. For every burst of magma researchers had seen at the surface, for every island that had been repaved, an equal and opposite effect might be taking place in the deep interior.

With these insights, it became that much more important for geophysicists to have a clear picture, even if it was just a crude cartoon, of how this boundary layer looks and how it gives rise to hot-spot plumes. To that end, researchers have often returned for guidance to the only terrain they have within their sights, the landscape at the surface. Just as they can envision landmasses similar to continents floating at the core-mantle boundary, they try to envision how plumes might form down there.

They know, for example, that plumes in the atmosphere form most commonly over mountain ridges. On the craggy heights of a mountain range the air is heated on a slope. Once it's warm, it runs up the sides of the mountain and to the peak, where it rises like an invisible smoke signal into the sky. Can some sort of similar topographic effect control the location of plumes in the anti-world? If so, then which are more important: the positions of anti-peaks and anti-valleys at the boundary or the insulating effects of anti-continents?

Different scenarios have different repercussions throughout the rest of the planet. The many possibilities have enabled geophysicists to dream up elaborate theories that connect plumes not only with features of the mantle and crust but also with extraordinary events that have taken place in the earth's long history, from changes in the magnetic field to biological extinctions. Some of these broad theories will show up in later chapters. But in rushing to get to these larger ideas, we should not overlook the other milestones that geophysicists have passed on their way down through the earth.

They put their images of plumes, the mantle and the boundary layer together and began to fathom the interconnectedness of the world's many parts. They saw their first glimpse of the earth as a complex system with checks and balances, inputs and outputs. Yet even after this breakthrough in thinking, geophysicists were still only halfway to their goal.

The boundary layer, in guiding the flow of heat, influences not only the convolutions of the rock above it but also the sphere below. Deeper and deeper into the earth, past the layers of rock, lies the raw metal of the core. Farther from us than Moscow is from the coast of France, it continuously broadcasts its presence through the invisible cloak of magnetism that it radiates around the planet.

Like a lighthouse beacon, this magnetic field has guided ocean voyagers for hundreds of years. By exploring that history, geophysicists would venture the next step of their journey of discovery. To do so, they joined forces with what had been until then an isolated group within their midst: the researchers known as geomagnetists.

6

MEANDERINGS OF THE MAGNETIC FIELD

So often, when calling on a geomagnetist anywhere, I have found a British accent answering my questions. David Gubbins at the University of Leeds offered an explanation for this phenomenon. "Historically, there were a lot of people in this country working on geomagnetism, probably dating from Lord [Patrick] Blackett, who started paleomagnetism" during the 1940s. Blackett turned out a handful of eminent students, and they then spawned a third generation in which the best and brightest young scientists were steered into geomagnetism. "All the chairs in the country are people of my age," says Gubbins. And many other geomagnetists have emigrated across the ocean to the United States.

Gubbins himself started under the tutelage of Sir Edward Bullard, one of Blackett's students, who became famous for developing the theory of how the earth's magnetic field is generated. This theory, published in the mid-1940s with Walter

Elsasser at Princeton University, was based on the idea that the field arises from the motions of molten iron in the outer core. Elsasser and Bullard proposed that electric currents in the iron flowing in the core would generate a magnetic field in much the same way that electricity running through a coil of copper wire creates a magnetic field around the wire. That physical phenomenon is the muscle behind a dynamo, or standard electric generator. Magnetic fields can produce electric currents and vice versa. So as long as the outer core was hot enough to be molten, Elsasser and Bullard's theory went, this dynamo at the center of the earth should persist, guiding travelers over land and sea.

That theory has survived the intervening decades pretty much intact. Modern researchers stand behind this broad explanation for the earth's magnetic field, and more to the point, they are hard-pressed to take it any further. "We don't know the answers to any of the really basic questions," says Coerte Voorhies, a geomagnetist at NASA's Goddard Space Flight Center. No one understands, for example, how the molten iron in the outer core flows or how that motion creates the magnetic field found on the surface. No one understands what causes the peculiarities of the field: the way its poles are constantly roving, the sudden jerks detected at observatories or the field's slow westward drift. These are the few clues that researchers have to the geodynamo's workings, yet they are no more intelligible than the mutterings of a lifelong hermit.

Ironically, geophysicists all agree on what's happening in the core mathematically. They can write down a set of some thirty coupled equations that describe the motion of the fluid. But as David Stevenson, a theoretical geophysicist at Caltech, once explained, "Even though we know the equations that should describe what is happening in the core, and everybody agrees that they're the right equations, the equations are incredibly difficult. They're more difficult, in fact, than the equations to describe weather in the earth's atmosphere. And you can't predict weather, even with the biggest computer available, beyond about four days ahead." Thus the geodynamo remains as one of the great mysteries of the earth. In 1905 Einstein ranked it among the top five unsolved problems in physics, and that status continues to this day.

Gubbins set out, over a dozen years ago, to explore one aspect of the dynamo: mapping the field in space. In particular, he wanted to know what the field looked like just as it was leaving the surface of the core. Tiny variations in the field's strength and direction could tell him something about the fluid flow—where the molten iron swirled into eddies, for example. And that could, in turn, shed some light on how the geodynamo works.

Gubbins knew he could accomplish this with a technique known as "downward continuing," in which one projects the surface field backward in space to show how it looked when it left the core. Many geomagnetists had tried and rejected this technique in the past because small errors in the measurements at the surface produced large errors at the core. Even now some researchers harrumph at the method. But Gubbins, along with colleagues at the Scripps Institution of Oceanography in La Jolla, California, developed a mathematical way to eliminate the largest aberrations. That breakthrough toppled any hurdles left in Gubbins's mind. He felt it was finally possible to create maps which, while lacking in some detail, would present a simple, streamlined view of the magnetic field at the core-mantle boundary.

At the time of the breakthrough, Jeremy Bloxham came to do a Ph.D. thesis in Gubbins's lab. Bloxham had taken his undergraduate degree in math at Cambridge, where Gubbins was then a professor, and he drifted into earth science just as Gubbins was poised to take a giant step. Armed with this new method of downward continuing, both researchers realized that they could make the greatest contribution by calculating how the magnetic field at the core-mantle boundary had evolved over time. That would show at least a hint of what the flow itself looked like. Gubbins set to work on the modern era, which left Bloxham to delve into the logs and files that had been left by the centuries-old tradition of ocean exploration.

Christopher Columbus paid close attention to the magnetic field swirling around his ships as he sailed west in 1492 toward what he thought was the coast of Asia. In the early evening, just as the stars were coming out, he would often pull out his quadrant and compass. The first he sighted on the North Star. The second he used to take its magnetic bearing, which he then scribbled into his daily log.

Columbus was one of a long line of European sailors dating back to the twelfth century who used the compass as their guide on cloudy nights. Chinese scientists and navigators had been taking detailed readings with the compass for a good 1,500 years before that. The magnetic field is thus one aspect of the earth that scientists have studied in detail for centuries. By 1700, elaborate charts showed the field's strength and direction at points all over the earth's surface.

Bloxham's task was to gather as much of the magnetic data as possible from the Age of Exploration. A great deal of that historical information was conveniently stored at the University of Edinburgh, three hundred miles northwest of Cambridge. Bloxham spent several months amid the dusty tomes of ancient voyages. Most of that time was spent on reams of navigational corrections. Since the mariners' crude instruments could give them only a ship's latitude, they had tremendous difficulty fixing their positions on the globe whenever they were out of sight of land. Yet each time they encountered an island or continent, Bloxham could correct their plotted positions and thus produce a more accurate magnetic reading. It was painstaking work correcting for tens of thousands of magnetic measurements, but the fact that Bloxham was himself a sailing enthusiast helped the months pass more quickly.

When the time finally came to turn this heap of records into maps of the field at the core-mantle boundary, Bloxham ended up relying heavily on a few key voyages. Between 1698 and 1700, the English astronomer Edmond Halley, of comet fame, sailed across the Atlantic on the *Paramore,* the first vessel to be commissioned for scientific work. Halley kept detailed records of the field as well as his location at sea, and his logs made up the bulk of Bloxham's first map, which was centered on the year 1715. For other maps Bloxham plumbed the records of explorers like Captain James Cook and Sir James Ross. All told, Bloxham produced three maps of the magnetic field, roughly fifty years apart, using records from 1695 to 1840.

In the meantime, Gubbins was working backward from the present, using modern data from magnetic observatories and from satellites like Magsat, which was launched in 1980. This modern data provided such a wealth of information that Gubbins charted maps at intervals of a decade instead of half a century. His projections of the surface field onto the core centered on the years 1980, 1970 and 1960.

Thus, working from both ends, the researchers produced a magnetic-field story line. The startling thing to emerge was that some of the features in the centuries-old maps showed up again in the modern ones. "There's no way that could happen by accident," says Gubbins. On the contrary, the common features gave the researchers a boost of faith. Bloxham and Gubbins could exult in the knowledge that, in broad strokes, these maps were accurate—not models of how the magnetic field at the core-mantle boundary *might* look, but working maps of the field as it was decades and centuries ago. Like the tomographers' coup in the mantle, Gubbins and Bloxham had stretched to the limit their ability to describe what was happening at the surface of the core.

In the years that followed, other graduate students in the lab filled in various gaps. Andy Jackson, now at Oxford University, came along and worked on the years between 1840 and 1900. Then Ken Hutcheson, now at the University of Glasgow, went even further back in time, gathering records from 1601 to 1695. As it happened, he found only about two thousand measurements in the entire seventeenth century. For this reason he and Gubbins decided it wouldn't pay to go back further.

In the meantime, Jackson spent a summer at Harvard University, where Bloxham had become a professor, turning their static maps into a series of movie frames. Gubbins added another decade's worth of modern satellite data to the film, and pretty soon the researchers had a motion picture. At first it was "a little Charlie Chaplinish," Bloxham says, but eventually it showed in a smooth continuum the magnetic field at the core surface from 1600 to the present day.

The film opens on a brightly colored globe on which the patterns of the shifting field are displayed. The globe is blue wherever the magnetic field lines dive inward, and thus the Northern Hemisphere is primarily blue, much as scientists might have guessed from the simple north-south orientation of the field at the earth's surface, as if a giant bar magnet were buried inside the planet. Similarly, the orange patches, where field lines emerge, cover the bulk of the Southern Hemisphere.

It is where the map diverges from this simple pattern, however, that researchers find meaning. Throughout the film, for example, four spots do not move much over the centuries. The field at these sites is particularly intense, as if the magnetic flux lines have been tied into tight bundles. These flux bundles lie at 60 degrees north and south along a line running through the Americas and another running through eastern Asia and Australia.

Based on what they knew about core movements, Bloxham and Gubbins speculated that these flux bundles marked the ends of tornado-like "rollers" or rotating columns of liquid in the outer core. Their theory harked back to an idea that Friedrich Busse at UCLA proposed in the early 1970s. Busse and his colleagues had built a hollow plastic sphere that they used to model the fluid motions of the core. They would fill the sphere with water or some other liquid and typically throw in a handful of fish scales to see the motions of the fluid inside. The fluid swirled around a smaller sphere that represented the solid inner core. And when they rotated the plastic model, the researchers found that rollers formed parallel to the axis of spin.

Busse then went on to show that, in theory, such rollers could give rise to the earth's magnetic field. This notion of how the geodynamo might work has lingered in the minds of geomagnetists. Although some now argue that this view of core circulation is too simplistic, Bloxham and Gubbins's maps appeared to lend some credibility to the idea.

The positions of these flux bundles led to an even more important connection. Three out of four of them appeared to be in spots where seismic-tomography maps showed patches of cold material in the lower mantle. Bloxham and Gubbins suggested that the geodynamo was somehow "locked to the mantle." Perhaps heat could pass more quickly out of the core in these cold patches, and by influencing how the molten iron flowed, convection in the mantle could thus be shaping the magnetic field. Regardless of the mechanism, this was the first time that geophysicists could see physical evidence of a connection between the mantle and core flow.

The tremendous breakthrough that came out of Bloxham and Gubbins's maps marked the beginning of the end of the isolation of two camps: investigators of the core and of the mantle. Dynamo theory had developed as a relatively isolated discipline within geophysics, and not just because it thrived in Great Britain. The geodynamo was a terribly difficult problem, and the handful of people working on it often seemed to their peers to be lost in a world of mathematical constructs that were either too complex or too esoteric to warrant other researchers' attention. But with Bloxham and Gubbins's maps, researchers began to see the iron flowing in the core as somehow slaved to the churning mantle above, and they even entertained the notion that the opposite might be true. "I certainly believe the mantle controls certain aspects of the core movements and the magnetic field," says Gubbins. "And it's quite conceivable that the core has a lot more control over the mantle than we normally think it has."

In this new light, geophysicists and geomagnetists found reason to pool their efforts. They started to view all of the earth's layers as inextricably intertwined. Twisting and turning, roiling and churning, the core and mantle were impinging on one another continually. Geophysicists and geomagnetists soon brought their combined efforts to bear upon yet another facet of the geodynamo: how and why the earth's magnetic field periodically reverses.

Looking at Flips

In the last 200 million years the field has reversed itself about three hundred times. As little as 30,000 years ago, the magnetic north and south poles were the opposite of where they are today. Human beings were roaming across most of the continents at that time. One night a group of Australian aborigines built a roaring fire and then left it to die on its own. As the red-hot rocks in the fireplace cooled, they adopted the orientation of the earth's magnetic field in much the way that molten rocks spewing from the mid-ocean ridges do, creating magnetic stripes on the seafloor. Centuries later, the stones in that ancient cooking fire were unearthed by a curious graduate student, providing the first piece of evidence for this most recent reversed-magnetic era.

Researchers have pieced together the history of a reversal from other sources. In a few rare cases, lava flows on land have taken place just as the magnetic field was undergoing a reversal. Sedimentary rocks can also capture the earth's magnetization in the alignment of mineral grains known as magnetite as they are laid down in sediments. From these occasional records, geophysicists have built up a picture of what it would be like to live through a magnetic reversal.

This question is of some importance because we may be headed for a reversal right now. Magnetized particles in Roman ceramic artifacts fired some two thousand years ago show that the earth's field then was about 75 percent stronger than it is today. One of the common traits of all magnetic reversals is that the intensity of the field gradually weakens over thousands of years to a fifth of its greatest strength. The field then reverses and builds its strength back up in the opposite direction. If the field continues to decay at its present rate, it will reach its nadir in some 1,200 years.

Until recently geophysicists thought that at this low point the magnetic field would also go haywire. There weren't many records to judge from, but what researchers did see convinced them that the weakened field of a reversal wouldn't point north and south like the "bar magnet" field that exists today. Instead the field lines would be going every which way. Two people standing at distant points on the globe would have completely different ideas about where the magnetic north pole lay.

For over a dozen years geomagnetists proceeded on this assumption. In gathering data they concentrated on records from individual sites because it was thought that only in this way could they gain any insight into the geodynamo. But when that insight never came, many researchers began to feel they were just treading water. Then in 1991, geomagnetist Brad Clement made a strange discovery.

Now at Florida International University, a few years ago Clement was a graduate student at Columbia University's Lamont-Doherty Earth Observatory. This research center, set on a scenic 125-acre site in the woods of Palisades, New York, houses the world's largest collection of cores extracted from the ocean floor. An entire warehouse is filled, floor to ceiling, with these dense plugs of mud and rock.

One plug in particular Clement was itching to examine: the first core to be drilled in the Southern Hemisphere in ocean crust dating back to the Matuyama-Brunhes magnetic reversal, which took place some 780,000 years ago. Dennis Kent, Clement's advisor at Lamont, had extracted the core from the southwestern Indian Ocean in 1985, and though Kent had studied the geochemistry of the core he had not yet analyzed the reversal data. Nor did he seem to Clement to be in any hurry to do so. "But I had seen them and I knew what the data were going to look like," says Clement. He thought the data were important, especially since there were no records of this reversal from the Southern Hemisphere.

Clement's urgency stemmed from the pattern he saw in that reversal. He wanted to expand the notion of what geomagnetists call the VGP, or virtual geomagnetic pole, which is the position of the north pole given by the magnetized particles in a rock sample. If, as geophysicists suspected, the field scrambled during a reversal and then reorganized itself in the opposite direction, the VGP for any one site wouldn't necessarily match the VGP at another. It would follow its own unique path from north to south, or vice versa.

Yet in the core from the Southern Hemisphere Clement saw the VGP path cutting a longitudinal swath right up through the Americas. This was the same path shown by a record of the same reversal from the opposite hemisphere. And when he looked at the other transition records for this reversal, he found that most of them either lined up along this American longitudinal path or one exactly opposite to it, passing through Asia and Australia.

That meant the pole was simply swinging from south to north along one of two direct paths. The field wasn't behaving like a simple bar magnet, since not one but two VGP paths were visible. But it also wasn't deteriorating into complete chaos. Clement had discovered a remarkable pattern that was unlike anything that geomagnetists had expected from the field. And though he felt strongly that what he was seeing was real, Clement realized that publishing it would also mean putting his reputation on the line. "I knew it was going to be controversial," he says. "Actually, I had some long talks about it with some colleagues of mine who I trust who work in the same area."

Key to these discussions was the fact that the positions of the reversal paths were strangely familiar. The longitudinal paths lay in the same places where Bloxham and Gubbins had seen their flux bundles. "I saw their maps at the same time that I was beginning to grapple with the results we were seeing," Clement says, "and I did, in all honesty, have those maps in my mind when I was examining the bulk of the data."

Encouraged by the correlation, Clement submitted his results to the prestigious journal *Nature.* However, the editors there rejected his paper. The reviewer, says Clement, had a knee-jerk reaction against it, calling the idea "pathological" and "outrageous." Only slightly deterred, Clement regrouped, rewrote the paper and submitted it to another journal, *Earth and Planetary Science Letters,* where it was published in 1991.

But by that time Clement's original paper had made its way around the geomagnetists' samizdat network and piqued the interest of a few seasoned pros. *Nature*'s editors began to sense they had made a mistake. At this time French geomagnetist Carlo Laj and his colleagues had uncovered reversal records that presented a bold, new twist on Clement's findings. And that is how it came to pass that only a week after Clement's paper was published—in effect, simultaneously—Laj reported similar findings in *Nature.* What's more, the journal granted that paper the coveted cover illustration, though it was only in the short "correspondence" section of the magazine.

The further step Laj and his colleagues took was to show these same magnetic pathways popping up repeatedly in reversals dating back at least 10 million years. This was the revelation that brought reversal pathways to center stage. Researchers simply could not ignore all those examples following the same paths and therefore couldn't escape the obvious conclusion.

The fluid in the core moves so rapidly—it is thought to have nearly the viscosity of water—that it should have no memory of past reversals, no way of harking back to the same longitudinal path time after time. Therefore the meanderings of the core field must be controlled or at least guided by the overlying mantle, which churns much more slowly. Hot and cold regions at the base of the mantle could be dictating the pattern of flow in the core, which would then influence the pattern of the magnetic field. It was one more unmistakable sign that the core and mantle are linked. To underline this, Laj's group published their reversal records alongside not only one of Bloxham and Gubbins's maps but a seismic-tomography map of the lowermost mantle. As they pointed out, the flux bundles also lined up with cold patches near the core-mantle boundary.

For geomagnetists, this odd discovery that the core has certain preferred pathways during reversals has provided a glimmer of hope that the geodynamo problem may not be as complicated as once was thought. At least during the special case of reversals, it may be steered by the mantle. And that gives dynamo theorists a way to solve one aspect of how the molten iron mixes.

For geophysicists in general, it is yet another chunk of evidence tipping the scales toward an integrated view of the earth. First seismology showed that currents in the mantle were not only connected to features at the earth's surface but to landmasses floating at the core-mantle boundary. Researchers began to see a billiard-ball pattern linking both of these features to hot-spot plumes. Then the maps of the magnetic field over history and, later, reversals of the field showed a similar parallel with this pattern, taking the coordinated motions of the interior hundreds of miles deeper, all the way into the earth's core.

Seeing all these connections in such a short span of time seems to have left researchers a bit dazed. "I just gave a seminar at my university today," geomagnetist Ken Hoffman said recently, "and I called it 'Earthquakes, CAT Scans and Geomagnetic Reversals.' And, you know, I had to say at some point that I never dreamed that I'd be able to try to marry these different things. And yet the correlations just knock my socks off. . . . You know, it gives you this impression that there really is kind of this holistic, perhaps simpler-than-ever-believed way of looking at the earth as a heat engine."

Hoffman has captured the sentiment of the times. Earth scientists now believe they are on their way to a second overarching synthesis. "The geological synthesis was the plate-tectonic revolution," says geomagnetist Mike Fuller, pondering the recent discoveries. "And now there's a deep interior revolution coming about."

Even after they have bubbled over with enthusiasm, though, most geophysicists will point out the many stumbling blocks that still stand in the way of this whole-earth revolution. Having known about geomagnetic reversals for over two decades, geomagnetists have only just discovered this powerful signal in the reversal record. Hoffman, whose own theory of the magnetic field's behavior during reversals was toppled by this new discovery, calls the situation "absolutely embarrassing." More to the point, it is a good reminder of the vast ignorance that reigns when it comes to the earth's core.

That ignorance is at the root of geophysicists' struggle with the larger problem of how the whole earth works. Like a distant planet, the core of the earth is still too far away and too unfamiliar for researchers to have made complete sense of its behavior. The temperatures and pressures inside it are so extreme that geophysicists don't completely understand how normal materials fare under these conditions. During the last decade, as discoveries about the deep earth have accumulated and revealed many of the connections in the earth engine, one key issue has stymied researchers' efforts to go deeper: Just how hot is it inside this central furnace?

7

THE BIGGEST GUN IN THE WEST

It is a short walk from the red tile roofs and stucco walls of Caltech's central campus through the meticulously kept gardens of the commons to the modern gray building where geophysics is done. Like so many other earth-science buildings around the country, this one has a glassed-in seismograph near its front door. The instrument's tiny pen clocks the motions of a patch of land several miles away, which means that no amount of jumping up and down next to this seismograph will disturb its steady line. But then neither will the violent shock waves that occasionally thunder up the stairwell.

For the basement of this building is home to a powerful cannon. The gun was pieced together over twenty years ago by Caltech geophysicist Tom Ahrens. He scavenged armaments from a navy destroyer and laid them end to end with fat cylinders of pipe to make a cannon that spans over 105 feet.

Yet this cannon is no weapon, and Ahrens doesn't use it to

take revenge on his fellow academics. He built the cannon to compress samples of minerals violently and instantaneously. For a fraction of a second, the gun is capable of generating temperatures of thousands of degrees Fahrenheit and pressures millions of times that at the surface of the earth. These are the extremes at the pit of the global centrifuge. The power to re-create them inspires a touch of hubris in geophysicists. It was this opportunity to peer into the planet's deepest recesses that lured Ahrens's colleague Jay Bass down to the gun room a few years ago to do an important experiment.

Every firing is the geological equivalent of a crash test. A pea-sized projectile is hurtled into a target at speeds of up to sixteen thousand miles an hour. In order to reach these speeds, five pounds of explosive propellant have to be loaded into the breech end of the gun. The propellant is a combination of nitroglycerine and nitrocellulose, and the trained technicians who handle it are careful to ground themselves so as not to raise any stray sparks that might set the gun off.

There have been accidents in the past, on this gun as well as the handful of others around the country. Ahrens's cannon blew apart during one early firing because he had used ordinary bolts to hold the sections of pipe together instead of bolts designed to withstand the strain. The explosion blew open all the doors and windows and ventilators in the gun room. "Had we not taken all the precautions we take," says Ahrens, "it would have certainly involved one or more fatalities." But the gun's operators, as in all the other incidents so far, had already sought cover.

"Once the powder is in there, you get away from there and you don't go near it," says Bass. "There's a concrete bunker with remote controls for operating the gun and most of the work is done from there." The operators run through a NASA-style checklist to make sure all the instruments are working properly. "And then the last thing you do, if everything checks out," says Bass, "is pull the safety cover off the fire button and hit it."

The explosives ignite, sending a fifty-pound piston whizzing down the cannon's barrel. The piston squeezes hydrogen gas in the barrel, which blows out a diaphragm separating the first stage of the gun from the second. (Technically the cannon is known as a two-stage light-gas gun.) Once that diaphragm bursts, all the energy of the giant piston is funneled into a narrow tube where the pea-sized projectile is waiting. Impelled to thousands of miles an hour, the projectile slams into the target at the end of the tube, a sample of earth minerals, and a shock wave rips through the sample and spreads throughout the gun room.

Immediately everyone in the building knows that the cannon has gone off, though not from the sound of the collision, which is a disappointing clunk. Bass notes, "Really, it makes a noise that sounds like you took a medium-sized rock and threw it into an empty garbage can. But all the upper floors of the building feel the shock, and you can feel it in the next building as well."

During the fall of 1986, these shock waves were rippling through the building frequently. Bass, Ahrens and Bob Svendson, then a graduate student at Caltech, were cowering in the concrete bunker, firing projectiles into samples of pure iron—poised to stake out new terrain in a historical controversy over a fundamental question: How hot is the center of the earth?

Bootstrapping

Such a basic question is a gaping hole in geophysicists' understanding of the planet and therefore a matter that they have returned to again and again. Most of what geophysicists know about the temperature inside the core comes from a technique known as bootstrapping. This is the intellectual equivalent of pulling yourself up by your bootstraps—it is using what you know to say something about what you don't know. One simple example of the process would be estimating the height of a child by looking at the height of the parents. It would be more accurate just to measure the child directly, but if you couldn't, the figure would probably be in the ballpark.

Astronomers bootstrap from nearby stars to more distant constellations to the farthest edges of the universe. Geophysicists bootstrap from what they know about the earth's near-surface layers to what they don't know about deeper ones. But with each additional bootstrap, errors accumulate and researchers can never really know how far afield their estimates are. Therefore their perceptions of the earth's temperature have long been based on a daisy chain of increasingly tenuous evidence.

A few facts they know positively. For instance, by digging boreholes into the ground geophysicists have learned that, roughly every hundred yards, the temperature rises 5 degrees Fahrenheit. That steady increase can take them several miles down into the crust.

At greater depths, researchers have found other bootstraps. Forty-five miles beneath the surface, the layer known as the asthenosphere begins. Geophysicists know the pressure on the rock at this depth. And because the layer is partially molten, they also know that a good estimate for the temperature 45 miles down is the melting temperature of rock at that pressure—roughly 2,000 degrees Fahrenheit.

Similarly, since the mantle rock is known to firm up again 155 miles below the surface, geophysicists have looked for the melting temperature at the corresponding, higher pressure. For the rock at this depth to stay solid, they believe it can be no hotter than about 3,000 degrees Fahrenheit.

In this way geophysicists have gauged the inner earth's temperature at several points on the way down. But they also know that these bootstraps can't take them very deep into the earth. That's because many radioactive elements, which produce heat, migrated toward the planet's surface during its early reorganization. As a result, the earth's outermost layers are inordinately hot. If the temperature of the earth continued to rise at the rate that it does in these outer layers, the core would be over 45,000 degrees Fahrenheit and the earth would be more like a puddle than a planet. But the radioactive elements thin out with depth, and so does the heat they generate.

What geophysicists have been lacking is a way to measure the temperature deep down, where radioactive heating plays a lesser role. That would show them exactly how temperatures arc through all the layers of the planet. And from that arc they could tease out clues to how the earth began as well as how it has churned out that heat throughout its long history.

For years they have had their sights on a single bootstrap. They have learned from seismology that the outer core is molten and the inner core is solid and that both layers are made of nearly pure iron. Nickel and traces of elements like sulfur, oxygen or silicon muddy the core a bit. But geophysicists knew that if they could figure out the melting point of pure iron under high core pressures, they would possess a very good estimate for the temperature at the slushy boundary between the inner and outer cores.

A half dozen years ago, the closest that geophysicists had come was to put the metal under 200,000 times atmospheric pressure—the force on the rock just 370 miles into the earth. From this melting temperature researchers extrapolated over 2,500 miles down to the core's liquid-solid boundary. Only with an instrument like Ahrens's cannon could they leap directly into the pressure range of the core. But even that cannon had to go through an initial stage of trial and error.

The technique of shock compression has actually been around since before World War II. Scientists working on the Manhattan Project used it to study how materials behaved when they were rapidly squeezed, since that was how they achieved the reactions for Fat Man and Little Boy. After the war, in 1958, researchers at Los Alamos compressed the first earth materials in the cannon. From the start, they could squeeze these materials to core pressures. And high temperatures automatically followed, since the whole process was nothing more than a large explosion, albeit a controlled one.

The experiment to measure core temperature, though, had to wait nearly thirty years. Geophysicists had to figure out how to detect the all-important transition from a solid to a liquid so that they could tell when the iron had melted. Not until the early 1980s did J. Michael Brown and Robert McQueen discover that they could tell when the iron had melted by looking at the speed of sound waves passing through it. Drawing on this newfound ability, geophysicists then went the distance into the earth's core.

Controlled Explosions

Bass and his colleagues proceeded through countdown after countdown; a red warning light continually flashed outside the gun room door. With the cannon they couldn't necessarily control what temperature they would achieve. Since every material has its own set of properties, the temperature of the shocked sample was simply a consequence of the pressure it endured on impact. So in order to reach the temperature where iron would melt, Bass says, "we beat on it harder and harder."

Each time the projectile screamed down the barrel of the cannon, it triggered a strobe camera that measured the thermal radiation emitted by the sample and told its temperature. After each dull clunk the researchers examined these traces. The sound-wave results had predicted that melting would take place around 2.5 million atmospheres. And at just that pressure the researchers saw a kink in their graphs where energy from the explosion was converting the solid metal into a liquid rather than raising its temperature. That was the proof they needed. The experiments had given them one point—the melting point of iron at a pressure well within the core.

But compared with past estimates, this one was a surprise. The melting point of iron was exorbitantly hot. Coincidentally, 350 miles up the coast at the University of California at Berkeley, another lab had independently arrived at a similar, higher-than-normal result. This lab was headed by Raymond Jeanloz, a bright light in the field of geophysics and also a former student of Ahrens's.

Jeanloz was at Caltech in the late 1970s, working on the cannon. Yet during those years, he says, "I became very sensitive to the idea that if you want to do basic physical chemistry, shock compression has one real problem." The results of a split-second reaction may not be the same as what would happen given the vast duration of the earth's history. "And that," says Jeanloz, "got me interested in the diamond anvil."

Like the cannon, the diamond anvil is used to mimic the extreme conditions inside the earth. But unlike the cannon, it can maintain those conditions indefinitely. Thus when Jeanloz set up his own lab at Berkeley, he centered it around this miniature vise. And by some accounts, the experiment to measure the melting temperature of iron was inevitable.

"It was pretty apparent that it was a project that had to be done somewhere along the line," says Quentin Williams, who arrived at the lab as a graduate student in 1983. But before they could attempt the experiment, Williams and another student had to put in a great deal of time calibrating the instrument. The sample squeezed inside the anvil is just a thousandth of an inch across, and the central bull's-eye that the laser heats is only a fifth that size, so using the laser caused wild fluctuations in the temperatures within the sample. From essentially ambient conditions at the outside edge, the temperature could rocket up to a peak of more than 9,000 degrees Fahrenheit, which made it difficult to tell exactly where along that steep ramp the sample was melting.

But as soon as they had completed their calibration, Williams and Jeanloz began squeezing samples of iron in the diamond anvil. This was just about the time that the Caltech group was shocking iron in the cannon. The Berkeley researchers took hundreds of different readings of the melting point, gradually increasing the pressure on the samples to over a million atmospheres—not quite core pressures, but more than twice as high as the next best set of anvil experiments. In most cases they could literally watch the tiny fleck of iron melt. "You can actually see that texture disappear and become a nice uniform little sheet of stuff," says Williams. "And you can see motion in the sample also."

Williams and Jeanloz then plotted the rising melting temperatures on a graph against the rising pressures on the sample. Like the path of an airplane during takeoff, the points on the graph rose steadily. When the researchers extended the line to pressures within the core, the results resembled those of the Caltech group. They pointed to core temperatures that were much higher than anyone else had ever seen.

At that point Ahrens suggested they join forces. The results from the cannon had the advantage of being at pressures well within the core. The results from the anvil, though they topped out at a lower pressure, reached that pressure by continuously sampling the melting temperatures all the way up the scale. Together they were a powerful duo. "There's always a tendency for people to discount one probe," says Williams. "But having two data sets which agreed very well made for a very strong story."

The paper came out in *Science* in April 1987, around the time that Williams was walking down the aisle with Elise Knittle. By anyone's measure, this was an eventful year. The researchers declared the temperature at the center of the earth to be approximately 12,000 degrees Fahrenheit. That was thousands of degrees higher than past estimates. In fact, it was hundreds of degrees hotter than the surface of the sun.

The Caltech and Berkeley groups knew their results were going to stir controversy, although it took a while to build, since the initial findings caught people by surprise. "And then people got a little more cranky about it," says Williams, laughing, "as they realized it might affect something they were doing."

If the high core-temperature results were correct, in the billions of years of evolution from a ball of galactic rubble to the lush orb of today the earth had retained a great deal more heat within its core than once was thought. A much hotter core meant that geophysicists would have to rethink what had happened in the rest of the planet as a result. The temperature gradient just above the core would become much steeper, for example, causing a much hotter boundary layer. That created problems for geophysicists trying to explain why the lowermost mantle wasn't molten.

The high core temperatures could also be used to support a theory that became fashionable among geophysicists during the 1980s. This is the idea that the moon was splashed out of the earth by the impact of a single giant planetesimal. The splashing theory accounts for a number of facts scientists have learned about the moon. It also implies that the earth was once a completely molten hunk of rock, which might explain how temperatures in the core could be so high.

Nevertheless, by any measure the high core-temperature results were radical. The breadth of their implications placed both the Caltech and Berkeley groups in the heart of a controversy. The inevitable scientific brouhaha proceeded, but this time it took on an angry timbre that no one would have predicted.

The Debate

"What's pathetic," says Jeanloz, "is that for the one other measurement that's lifted out of the literature and exhibited to us, exhibit A, showing that maybe there's something questionable about our results, the guy has yet to do a single calibration now even five years later."

Bass is just as adamant on this issue. "That may sound like

a technical point. But it's like having a speedometer which you just picked up at K Mart. So some little kid puts it on a bicycle and he doesn't hook it up exactly the right way according to the instructions, and he comes home and he says, 'Gee, I rode my bicycle a hundred miles an hour today!' And you've got to see, well, is this instrument any good, is it hooked up properly, or does it need to be corrected?"

Bass and Jeanloz's criticism is leveled at the research of Reinhard Boehler at the Max Planck Institut für Chemie in Mainz, Germany. If the Caltech-Berkeley group's paper represents one extreme, Boehler's results represent another. Just a year before the *Science* paper came out, Boehler had published a remarkably low temperature estimate for the center of the earth of some 6,500 degrees Fahrenheit.

He had performed his experiments in a diamond anvil similar to the one that Jeanloz and Williams had used, though he had employed slightly different methods for detecting the melting point of the iron and calculating its temperature. Such techniques can make all the difference in the outcome of an experiment. Many geophysicists harbored the notion that this was the cause of the temperature discrepancy. Nonetheless, Boehler's methods proved consistent.

By 1993, he had doubled the pressures that Williams and Jeanloz achieved in the diamond anvil and still found the same low core temperatures that he had predicted earlier. As he had done in the past, Boehler dismissed the California group's high temperatures, attributing them to large systematic errors in their experiments, much the way they had discounted his own cooler results.

That the debate between these two camps has been so caustic has much to do with the characters of its practitioners. Like Jeanloz, Boehler is a charming and gregarious fellow. But those who have worked with him concede that he is both strong-willed and temperamental. By the same token, Jeanloz's vibrant personality also has its fiery side.

Yet the fact that personalities have played such a role is largely a product of the question that these researchers are asking. When geophysicists are reaching as far as they can reach, trying to decide whose results to trust can be the most difficult task of all. It's as if each of these groups of researchers were trying to identify an object floating on a distant horizon. Is that object a sailboat or a freighter? At a great enough range it is impossible to tell. The best anyone could do under the circumstances would be to squint and make the best guess possible, which is what both groups of geophysicists are doing on this question.

In the end they are only human. As Tom Ahrens once put it, "Sometimes there's nothing wrong or right about things." People make their best effort to get the correct answer and they publish all the details of how they arrived at that answer. But there may be some problem, which they don't recognize, that is obscuring the truth.

Current theory has it that something is happening to iron at core pressures that the early experiments didn't take into account. Boehler and other researchers have proposed that the crystals within the iron may realign themselves under these extreme conditions, much the way that carbon molecules realign themselves when graphite turns into diamond. High-pressure researchers are now using the anvil and the cannon to search for signs of these internal changes. In the meantime, the geophysics community has pressed on toward a comprehensive theory of the earth in spite of such gaping holes in their knowledge.

8

LEAKS IN THE MACHINE

The image that comes to mind to describe the situation that geophysicists now find themselves in is the bronze-plated statue of Atlas that stands in front of the International Building in New York's Rockefeller Center. This hulking Atlas is carrying the universe upon his shoulders, a hollow orb ringed with the constellations of the celestial sphere. It seems that the orb geophysicists are carrying is similar. Instead of the heavens, it is the weight of the world they hold in their arms—a world that is somewhat fragmented. The many connections that researchers have forged between its layers in the last decade are just thin wires holding the sundry parts of the globe together. Thus geophysicists are having a terrible time maintaining their grip upon the planet, because the world as they see it is still all in pieces.

Geophysicists talk of a revolution in progress. But if this spate of discoveries is indeed to be the sequel to the plate tectonics revolution, the key to moving forward is to understand how the pieces work together. What geophysicists need is a blueprint for how the earth machine works.

The broadest strokes of that blueprint are already in place. Researchers know that the wheels within wheels of convection are driven by the earth's efforts to rid itself of heat. Its mammoth stores of rocks, all at extremely high temperatures and pressures, are nudging and squirting past each other to carry their parcels of heat to the surface and allow cooled parcels to drop back into the interior. But in addition to not really understanding how much the core is contributing to this heat engine, geophysicists face another quandary when they turn to the earth's mantle.

This is arguably the most important part in the earth machine. It mediates between the swirling metal of the core and the gliding plates of the surface. Thus, understanding how it stirs is one of the best ways for geophysicists to create an all-inclusive picture. In recent years, seismic tomography has brought researchers to the brink of actually seeing inside this layer, with its glorious orange and blue maps of the mantle's steady churning. Yet those who would explore the mantle have also had to face the fact that these maps simply do not go far enough.

Seismic tomographers have known all along that their maps would provide only an instantaneous snapshot of convection, but they had hoped that tomography would portray the mantle's style of motion, like a still photo showing the gait of a galloping horse. They were disappointed to find that the seismic snapshots were too blurry to support those kinds of mental leaps.

Says tomographer Robert Clayton, "If you just take a photograph of something easily recognizable optically, say a human face, you can degrade that thing to a tremendous level and yet people can still see the image of the face in it. What's happening is that your eye is picking up the patterns in it. We're looking for patterns in these tomographic maps, but we don't know what patterns we're looking for."

Or, more precisely, researchers are looking for two mutually exclusive patterns. Depending on which camp the scientists belong to, they can detect evidence for either configuration in these maps. The two patterns represent the two sides of a debate that has split the geophysics community for some twenty years: whole-mantle versus layered-mantle convection. That is, does the flow in the mantle extend from the crust to the core in a kind of grand, looping flow, or is there some sort of natural boundary around the transition zone that makes the mantle convect in two separate layers, passing heat but no material between them? The solution is the crux of the most basic question in geophysics, how the motions of the interior drive the contortions of the surface.

A Layered Layer

The two sides of the debate have traditionally divided along the lines of two branches of earth science, geochemistry and geophysics, which view the mantle in very different ways. Geochemists tend to focus on the chemical differences between the rocks that are extruded at mid-ocean ridges and hot-spot island volcanoes. That rock carries with it traces of the deep-earth layers that it passed through on its journey. And thus geochemists can use it to make a kind of chemical map of the mantle.

But unlike seismic-tomography maps, which provide only a frozen image of the earth's interior, geochemistry can study the changes in the mantle over time. That's because the rock that comes to the surface contains faint traces of radioactive isotopes, chemical elements that decay at a known rate. Detecting these trace elements in the rock of the earth has never been easy. The analyses must be done in special "clean labs" where the environment is well filtered. But by looking at the levels of various isotopes, geochemists can group the rocks that have similar deep-earth origins. Then they can use the amount of radioactive decay to calculate how long these groups, or "reservoirs" of mantle rock, have been isolated from one another. In this way geochemists have managed not only to map the different reservoirs of the mantle, but to gain some insight into how they have moved and changed over the eons.

In the 1970s, for example, University of Rhode Island geochemist Jean-Guy Schilling dredged up nine thousand pounds of rock from around Iceland, a unique spot on earth where both a mid-ocean ridge and a hot-spot volcano overlap. As Schilling moved sampling farther and farther away from the hot spot to the ridge alone, he noticed certain elements disappearing from the rock. The fact that rocks from the ridge and from the hot spot had different compositions implied to Schilling that they might also have different origins. Ridge and hot-spot magmas might be rising out of separate reservoirs in the mantle. In addition, the radioactive-decay rates of the dredged-up rocks implied that these reservoirs had been isolated from each other for at least a billion years.

The key finding, though, was that the very elements missing from the ridge rock were present in abundance in continental crust. That convinced geochemists that the reservoir of rock supplying magma to the ridges had given up its trace elements to the continental crust, probably early in the earth's history, and had remained isolated ever since. When researchers calculated the size of this reservoir, they found it to be about a third of the volume of the entire mantle, or exactly the size of the upper mantle. And so this elaborate set of circumstances laid the foundation for the argument that was widely used for two decades: that something had prevented the upper mantle from mixing with the lower throughout much of the earth's four and a half billion years.

That something was long thought to be a slight difference in composition, based on the widespread notion that the lower mantle contained a dash more iron than the upper mantle. If this iron increased the density of lower mantle rock, as was expected, then at the four-hundred-mile-deep boundary between the mantle's two layers subducting slabs of crust would bend and deform like wedges of butter rather than piercing through.

Throughout the 1980s, Caltech's Don Anderson was the main proponent of this layered view of the earth. He spent much of that time slowly accumulating evidence and arguments for his cause, not just from geochemistry but from all branches of the earth sciences. In this respect Anderson is a rare breed among geophysicists, an avowed generalist. He began his career as most geologists do, delving into rocks along America's eastern seaboard with his hammer and chisel. Later he took a turn looking for underground oil fields with seismic waves before eventually plunging headlong into studies of the depths of the planet.

With this wealth of experience behind him, Anderson feels entitled to a certain stubbornness in his beliefs. He has probed this problem thoroughly, and colleagues who spar with him admit that, for every argument, he has at least a plausible counterargument. "I'm convinced by the evidence that says the mantle isn't convecting as a whole," Anderson affirmed in a recent interview, "that slabs don't go into the lower mantle, and plumes don't originate in the lower mantle, that all the action is in the upper mantle except for the fact that the lower mantle is convecting."

An Alternative View

It is all the more surprising, therefore, that one of the most persuasive arguments for the other side of the mantle debate came from a former student of Anderson's, Tom Jordan. Jordan is now head of MIT's earth science department, but he studied at the California Institute of Technology in the late 1960s and then went on to graduate school there under Don Anderson's tutelage. The papers Jordan published while he was at Caltech reflected the influence of Anderson's layered-world view. But in 1972 Jordan left the institute, and just two years later he had embraced the view of the whole-mantle camp.

Starting in 1974, in collaboration with other researchers, Jordan published a series of papers showing that the slabs of ocean crust diving beneath first the Caribbean Sea and then the Sea of Okhotsk in the northwest Pacific were punching into the lower mantle. He used a shallow, regional form of seismic tomography to plot this, adopting earthquakes from within the subducting slabs to look for variations in the rock around them. And if anything could punch through the boundary, it would be these slabs, which, having cooled on the surface for so long, are some of the densest, most rigid rocks on earth.

Based on these results, Jordan began arguing for a less segregated view of the mantle. But his ideas didn't really catch on until the mid-1980s, when he and his student Kenneth Creager, now at the University of Washington, showed deep traces of slabs plunging down at least six hundred miles all along the western Pacific Rim, from the Bering Sea through Japan to the island of Guam. The researchers pointed out that if the situation was similar at all ocean trenches, then they could calculate just how much upper-mantle rock was going into the lower mantle. They demonstrated that it would take about a billion years at most to intermix the mantle and, therefore, a completely isolated upper and lower mantle just didn't fit their view of the earth.

It was an impressive body of work and strong evidence that the strictly layered model favored by Anderson and the geochemists was flawed. In conference sessions throughout the late 1980s, a civilized but spirited feud played out between these two camps. For most of that time it seemed that the majority of researchers were siding with Jordan's wholly mixed mantle. But more recently his results seems to have lost some of their luster.

Some researchers see the slabs fattening and deflecting around the boundary, as if they run into a wall. Others see slabs punching through in some places they look but not in others. And no one can prove that what appears to be a slab in the lower mantle isn't in fact just a seismic mirage caused by the slab's "coolness" spreading into the rock below the boundary.

With each passing year, seismologists have struggled to etch more detail into the picture of this critical boundary, expanding into new and creative uses of their craft. In a recent twist on the usual seismic techniques, Peter Shearer and T. Guy Masters of the Scripps Institution of Oceanography in La Jolla, California, used seismic waves to map the undulations of the boundary between the upper and lower mantle. Their results showed broad valleys where slabs of crust were shoving the boundary down, much the way a swimmer creates a wave in the water in front of him. Shearer and Masters felt these depressions would be much sharper if the slabs were actually poking through, and they have suggested that the crust is breaking up or bending at the boundary rather than punching through it. But rather than providing a solution, their results just added more fuel to the fire.

It's as if geophysicists are caught in a family feud. So much time has been taken up by the argument itself and with every incremental step forward, researchers drag along all the baggage of previous debates. For this reason, a number of geophysicists have begun to seek a different approach to this problem, to find a middle ground.

Our World Leaks

Caltech's David Stevenson summed up the views of many of his colleagues on this subject.

> When you see a conflict like that, there are two possibilities. Either one body of knowledge is flawed somehow—the geochemists are wrong or the geophysicists are wrong, to put it in a crass way. The other possibility is that in some sense they're both right, and the earth has somehow evolved into a state where in some respects it behaves like a layered system and in others like a whole-mantle convecting system. I'm inclined to the latter, not because I have one-hundred-percent confidence in either the geochemists or the geophysicists but simply because both sets of arguments do seem to me to be fairly strong. Somehow it seems as though material can probably get through some of the time but not all of the time. So it's not a true whole-mantle convection system, but a leaky system.

Our world leaks. In the simplest, cartoon view, what Stevenson and other geophysicists envision beneath their feet is not a single layer of mantle churning alone, nor is it two layers churning separately; rather, it is a pervious combination of the two—a rapidly mixing upper layer inside a slowly roiling, mantle-wide layer. Like cream on top of oatmeal, these layers have resisted homogenization throughout their long history. Yet they have intermingled slightly as they stirred.

Geochemists also are beginning to support this curious picture of the earth. For Stan Hart, who has worked on the geochemistry of the earth's mantle throughout his distinguished career, first at MIT and then at Woods Hole, the strictly layered view of the mantle began to fail in the mid-1980s. That was when Hart and his colleagues recalculated the size of the reservoir that geochemists had previously thought to be about a third of the mantle. Because of the improvements in sampling and measurement techniques, the reservoir turned out to be more than half the mantle, by their estimate. Simply dividing the mantle into upper and lower layers would no longer do—the earth was more complicated.

As the same time, the total number of reservoirs that geochemists attributed to the mantle was expanding. Over the years researchers had found even more distinctions between the chemistry of ocean island rocks, and by 1986 Hart was already arguing that the mantle must have not two but at least four separate reservoirs which had remained intact for over a billion years. Hart and his colleagues considered the upper and lower mantles to be the two main reservoirs. But somewhere in the midst of these layers two additional reservoirs had to exist. These might consist of old ocean crust or material from one of the boundary layers. Hart and his colleagues suggested that plumes originated from one reservoir and picked up material from the others as they rose through the mantle. With this explanation for their results, the geochemists had adopted the geophysicists' leaky view of the mantle.

The Big Flush

Having embraced the leaky idea, researchers were recently surprised by a batch of new computer models that provided them with a glimpse of how a leaky earth might actually work. Although modeling mantle convection is still an immature art, it has come a long way since the early 1980s, when even the best simulations of mantle convection were still being done in two dimensions, because of computer limitations. Longtime UCLA modeler Gerry Schubert says, "I believe that even though we realized the earth was three-dimensional, seeing presentations, figures, slides of largely two-dimensional convection cells for periods of ten to fifteen years had some effect on our perception of mantle convection."

Long ago, Socrates drew a similar conclusion when he described (and Plato later recorded) a group of people who could only see shadows cast by firelight onto the walls of a cave. Everything in their world was two-dimensional. And thus, Socrates argued, only when they left the cave, and even then with some difficulty, would they recognize the human forms behind their own gray silhouettes.

Only in the last few years, with the proliferation of supercomputers, have mantle modelers made that jump to three dimensions, and the results have been startling. Initially they looked at the way convection behaved in a cube. Then in 1989, Schubert and his colleagues Dave Bercovici, now at the University of Hawaii, and Gary Glatzmaier, at the Los Alamos National Laboratory, generated the first three-dimensional model of mantle convection in a sphere. The model showed researchers that the shape of the mantle's convecting currents—the narrow sheets of rock diving into the earth and columns of rock ascending—is not a side effect of having the rigid crustal plates on the surface. Instead, it is a natural consequence of convection in a sphere.

Then in 1993, Paul Tackley, a graduate student at Caltech, along with Schubert, Glatzmaier and David Stevenson, took the spherical model one step further. Along with some other improvements, they added what's known as a phase change. A phase change takes place, for example, when water turns into ice. When graphite, the material in pencil lead, is squeezed hard enough, it changes into diamond. Basically any material that is exposed to extreme pressures or extreme temperatures will no longer be content in its normal condition and will undergo a phase change—either by melting, evaporating, solidifying or collapsing from a less dense solid into a more dense solid.

Rocks in the earth's mantle undergo this latter kind of phase change at a depth of 250 miles and then again just past 400 miles at the bottom of the transition zone. Geophysicists have long known that these phase changes, particularly the one at 400 miles, might influence the way rock in the mantle mixes. But until recently the changes were largely ignored.

The reason for this stems from some simulations done in the mid-1980s. Those early models, which included phase changes, implied that any effect on mantle mixing was unlikely, particularly when the conditions in the model were as "earth-like" as possible. So geophysicists concentrated on whether a chemical difference between the upper and lower mantle was causing layering. But as Tackley points out, "They were just little, two-dimensional, box simulations."

Now that geophysicists have taken another look at phase

changes, they can come closer to simulating the extreme conditions inside the earth. High-pressure researchers have also provided a much more accurate description of the mantle's phase changes. In particular, they recently showed that each change takes place over a narrow band of a few miles at most, rather than ten or more miles, as researchers had once thought. What modelers found when they added these new results to their improved simulations took them by surprise.

It seems that inside the mantle, an unusual sort of "flushing" is going on. Across most of the mantle, the phase change acts as a barrier, halting the downward fall of the sinking slabs and causing them to gradually pool just above the 400-mile boundary. Thus, without any change in the composition across the boundary, the phase change produces a temporarily layered mantle. But then there are these avalanches. Every so often the pooled rock will build up enough mass that it breaks through the phase boundary in one patch. It flushes violently into the lower mantle, plunging all the way down to the core-mantle boundary despite the fact that the rock it's pushing through gets more and more viscous with depth.

Tackley's model shows a few of these flushes opening the floodgates between the mantle's layers. A number of other modelers have also seen flushing in both 2-D cross sections and 3-D cubes. (In a prankish variation on this sudden spurt, some modelers have even adjusted their computers to make a "flushing" sound effect.) From these unusual models comes the interesting possibility that the mantle has been both layered and mixed throughout the earth's history. Yet the real test of any model is whether its predictions take place in the earth.

So far the best evidence for mantle flushing has come from Dave Yuen at the University of Minnesota and his colleagues. They think they see signs of flushing in seismic-tomography maps. As Yuen points out, if slabs of crust were continuously slipping down into the lower mantle instead of flushing periodically, the tomographic images should show just as much cold material at the top, middle and bottom of that lower layer. But if flushing really does occur, then large patches of cold rock should be distributed more sporadically throughout the mantle's lower depths.

Recently Yuen and his colleagues have shown that this is indeed the case. "We found that there's a gap," Yuen explains. "So that means there must be a period of flushing, because if you have a steady-state model you would not expect to have a gap." They found the greatest concentration of cold patches just above the core-mantle boundary, around 1,800 miles deep. Very little cold material lies in the middle of the lower mantle between 840 and 1,550 miles depth. Then another, smaller concentration shows up near the top of the lower mantle, around 745 miles down.

It is still too soon to say whether these models are believable. As far as the models have come, they have not yet managed to mimic the earth satisfactorily. That we don't know such things as how much heat is being emitted by the core contributes to geophysicists' uncertainty about the computer models. But the next hurdle will be putting realistic plates on the surface. That means making the plates rigid, the way rock is when it's cold, in contrast to the hotter, more fluid rock deep in the earth. The behavior of plates is such a major consideration that many researchers believe it could have just as dramatic an effect on the theories of convection as the addition of the phase change did.

Geophysicists believe this piece will fall into place in the next few years—a sign of how quickly the field of modeling is evolving. Likewise, the techniques of seismic tomography are constantly being improved, and more and more digital seismometers are being set up in uncovered parts of the globe.

Plummeting oceanic slabs and even the skinny tails of hot-spot plumes may someday stand out in sharp definition, and everyone's doubts about the nature of mantle convection will dissolve into the ether.

Until that day, or until the models are more complete, those arguing for either whole or layered convection are standing their ground. Tom Jordan, still firmly in the camp of whole-mantle flow, says he thinks the avalanches have been overemphasized. "The seismic data suggest that if the earth has features like this, they're much smaller than these models predict, much less dramatic than these models predict."

Don Anderson, still championing layered convection, also points to the missing pieces in the avalanche models. "The problem with all these calculations is they assume a homogenous chemical mantle. But if there's even a slight difference between the material in the upper and lower mantle, these avalanches might never happen."

For the majority of geophysicists, however, the latest results have provided them with the valuable insight that the earth can be leaky over time as well as leaky in select places. Thus it has become the de facto model for how the mantle behaves.

If it is indeed correct, then it should fit with the other parts of the earth engine. But in particular, it should mesh with the planet's crust. On the surface, continents and oceans have existed for millions of years as the motions of the plates have continuously rearranged them. Guided by the churning within the planet, the crust of the earth has been in a constant state of flux. These motions are also the one aspect of the earth that researchers can see directly. Uniting them with the contortions of the interior is thus, to geophysicists, the backbone of any universal theory for how all the pieces of the planet work together.

9

THE CRACK-UP

Imagine the world as it was 200 million years ago. At that time there were no scattered continents, no Australia or North America. All the land was assembled into a single, giant supercontinent known as Pangaea. This vast crescent of land stretched from pole to pole and hovered roughly where the African continent is today. Along its jagged shores broke the waves of a worldwide ocean, Panthalassa. Within its sheltered arc lapped the waters of the narrow Tethys Sea. Moist ocean breezes rarely penetrated the supercontinent's remote central plains. For tens of millions of years, inland Pangaea spread out like an infinite desert, home to the numerous species of crawling reptiles that could survive within this arid terrain.

But the earth's machinations were soon to change all that. A hundred and eighty million years ago, the face of the world took a dramatic turn when the giant Pangaean supercontinent began to break up. From west to east a groove across its center stretched thin like pulled taffy. The eastern shore of what is now North America began to pull away from the northwest corner of Africa. Between them formed several long, narrow lakes and then, as the rift deepened, a wide ocean channel.

At last the sluice gates were lowered. Balmy, near-equatorial currents from Panthalassa rushed between the sundered continents along the newly formed Tethyan Seaway. Like the modern-day Gulf Stream, these waters warmed and humidified the two landmasses. The sweltry climate stoked plant life.

Then some 150 million years ago, the northern and southern sections of Pangaea also started to break apart. North America slowly detached from the west coast of Europe, and, in two decisive blows, first the Antarctic continent and then South America split away from the coasts of Africa.

The great Pangaean landmass was now riddled with miles and miles of coastline, and the world was slowly transformed. Lush swamplands swelled. Palm trees grew at high latitudes. The earth saw its first flowering plants and then a singular explosion of them as the barren plains burst into bloom. The family of dinosaurs, faced with this vast new range of habitats, rose to greatness by exploiting it. They advanced down the evolutionary paths that culminated in flying pterosaurs, plant-eating apatosaurs and the rapacious *Tyrannosaurus rex*. Pangaea's crack-up was one of the most phenomenal changes to the surface of the planet in geological history. Like some terrestrial Big Bang, it sent the continents reeling, and they have continued to drift outward ever since.

Geophysicists have known for many years that, like most changes to the surface of the earth, this cataclysm was a reflection of what was going on in the body beneath. Using clues left in the rocks, researchers in the 1970s reconstructed the breakup of Pangaea and the opening of the Atlantic Ocean. The same magnetic stripes on the ocean floor that first convinced geologists the crust was mobile allowed them to trace the motions of the plates backward in time. Several dozen hot spots, fixed in their positions like buoys in a bay, also helped researchers orient each moving plate on the surface of the globe.

Delving into the past, the researchers arrived at that distant era when all the world's continents nestled shoulder to shoulder. And along the crooked border where the landmasses once came together, the researchers made an extraordinary discovery. The hot-spot plumes that are now scattered around the Atlantic Ocean were all lined up like a connect-the-dots puzzle beneath the jagged Pangaean rift.

One such plume in the South Atlantic now lies beneath a warm, windswept island group known as Tristan da Cunha. On Tristan, the largest and only inhabited island in the group, some three hundred people, mostly descendants of shipwrecked sailors, make their living fishing and raising potatoes on the island's high plateaus. There is just one drawback to their island paradise: every so often it bursts into flame beneath them. Tristan last blew its stack in 1961, forcing a complete evacuation. But the hot-spot plume that created this island has been erupting in this way since long before the Tristanians came.

A hundred and thirty million years ago, the Tristan Plume was directly beneath the rift that split apart Africa and South America. Its bulbous head had flattened into an enormous pancake of magma beneath the supercontinent, and the heat caused the crust above it to swell into a dome. Then just before the continent broke into two, Tristan erupted onto the surface, leaving a massive flood of lava. This tremendous outpouring can still be found upon the continents on both sides of the break. In Namibia, the Etendeka basalt plain amounts to nearly 60,000 cubic miles; its hard, craggy rocks reach right to the edge of Africa's southwestern coast. Around São Paulo,

Brazil, the Serra Geral basalts are roughly five times that volume, even after substantial erosion, though the hilly terrain is now so densely forested that the dusky crags of this ancient eruption are no longer exposed.

Shortly after this flood of rock, the Atlantic began to open. The Tristan Plume lay directly beneath the mid-ocean ridge, carving a V-shaped line of volcanic islands into the ocean plates as they spread away from it. On the South American Plate, this submerged chain of seamounts is known as the Rio Grande Ridge. On the African Plate it is called the Walvis Ridge. They lie like a trail of bread crumbs on the bottom of the ocean, directing us back to where the eruptions began.

The picture is similar for other hot spots in and around the Atlantic. The plumes that created the Azores, St. Helena, Cape Verde, Trinidad and the Madeira Islands are all part of the great necklace of hot spots that once burned beneath Pangaea like so many signal flares. The dozen-odd plumes now feed volcanoes on the African Plate, part of the deep upwelling of hot mantle rock that is splitting that continent apart today.

For some 200 million years, then, this hot region of the mantle has been perforating the surface. This raises the question, did the plumes cause the Pangaean crust to fracture? A minor skirmish over this idea had been waged in recent years—a kind of chicken-and-egg argument about which comes first, the plume or the rift.

Two researchers at Cambridge University in England, Robert White and Dan McKenzie, have championed the idea that plumes play a more passive role in rifting. They believe that the existing stresses of tectonic motion stretch and thin the crust first, thereby allowing magma to rise and then flood the surface. As a result of White and McKenzie's influence, some researchers believe that plumes only dictate where the rift will be and tectonic stresses do the rest. But for the vast majority, plumes still come first; to do something as dramatic as rip a landmass in two requires more than the everyday shove and pull of tectonic motion.

Yet the geophysicists also believe it is a mistake to think of the surface of the earth as merely rolling with punches thrown by the interior. As much as hot-spot plumes have an impact on the surface, the same is true in reverse. Geophysicists have long understood that subducting slabs of ocean crust play a major role in how the surface and the interior mix together, even though what happens to the slabs has long been a matter of contention. Now, at a time when geophysicists are prepared to think about subducting slabs leaking through to the lower mantle, the idea has received a boost from some recent work by geophysicist Mark Richards at the University of California at Berkeley.

In 1983, Richards was at Caltech when Adam Dziewonski came for a visit with his brand-new tomography maps in hand. Before those models even made it into print, Richards was part of the team who showed that the blobs of hot and cold rock in the lower mantle matched the billiard-ball pattern of gravity at the planet's surface. "Shortly after that," he says, "I began to realize that you could build a pretty good model of mantle heterogeneity just by dropping slabs into the mantle." Thus began his efforts to re-create the history of subducting plates to see how much they had shaped the present-day earth.

The biggest problem Richards faced was finding someone who could reconstruct the plate motions with enough fidelity so that he could believe the results. He found those qualities in paleomagnetist David Engebretson at Western Washington University, who for some fifteen years has been using the magnetic strips on the seafloor to look at past plate locations. Yet even with that experience he found Richards's request a challenge. Just how much of the crust is diving into the earth at a given time is determined by the size of the plates' borders. These borders continuously shift like the edges of soap bubbles. Reconstructing plate motions is thus a process of nudging the plate fragments this way and that, trying to fit those fragments into a coherent picture at each successive step backward in time.

Even so, with Engebretson's maps and Richard's modeling skills, the researchers eventually arrived at a satisfying result. In its prime, Pangaea was surrounded by subduction zones. During the nearly 200 million years since the supercontinent reigned on the surface, that ancient circle of sinking crust has migrated across the globe to the Pacific, to form the arc we now know as the Ring of Fire. The next step was to consider the earth's interior.

Theoretically, it takes over a hundred million years for the cool aura of the slabs to pass from the surface to the core. When Engebretson and Richards dropped slabs from as far back as Pangaean times into the mantle and allowed them to slide through the boundary layer, they found that their models of where subducting slabs should be exactly matched the areas where tomography maps showed the mantle should be cool, right down to the core-mantle boundary. "The model is so good that it almost makes me worry," says Richards. "I mean, our model is as well correlated with any of the global models of seismic tomography as those models are with themselves."

Now, some researchers would argue that the slabs are not sliding though the boundary, that they simply have a cooling effect on the rock around them or that they punch through in sudden avalanches. But regardless of how the cold rock leaked into the lower mantle, the downwellings that are now quenching warm spots on the core-mantle boundary like water on burning coals are the traces of the great Pangaean landmass.

Today no hot spots bloom in that part of the mantle where the 200-million-year-old slabs were dumped. Even the tiny slab of ocean crust that lay beneath the Tethys Sea before that waterway was pinched off has contributed its part. Seismologists believe they can now see it chilling a patch of the core-mantle boundary beneath Australia.

This means that both the surface and the core-mantle boundary play an important role in the evolving machine of the earth. Sinking slabs of crust can dictate where plumes will and will not rise off the boundary. And the record of Pangaea's breakup surpasses even the modern-day rifting in eastern Africa as a testament to the enormous effects plumes can have on the surface.

Researchers are beginning to see a larger picture. The major events of the planet's history seem to stem from what geophysicists generally think of as the leaks in the machine. Plumes and sinking slabs of crust are part and parcel of the earth's grand scale of convection, one so broad that it links the shores of Japan on the Pacific Rim to the top of the molten core.

Over the breadth of history, researchers can see that events that took place eons ago are shaping both the present-day landscape and the eddies of molten iron in the core. By extension, the area where slabs are being dumped into mantle right now is dictating where clusters of hot spots will be able to rise from the core-mantle boundary some 200 million years into the future and pierce continents at the surface.

The Chaotic Plates

Yet the picture grows even more complex. Hot-spot plumes may rise and break a continent in two, starting a new mid-ocean ridge, but another, faster-moving mechanism in the earth machine takes over once that tear in the crust is complete. This is the flux in the shallow mantle, a steady turnover of upwellings at ridges and downwelling at trenches.

The motions of the plates are married to this shallow turmoil. Once hot spots and rifting have created a new border, the plates on either side of it start moving freely. That's because the mid-ocean ridges don't need to stay perched on top of the plumes in order to keep churning out ocean crust. In fact, it would be impossible for a ridge to stay in any one position permanently. With several ridges expanding simultaneously on the global sphere, basic geometry doesn't allow it. The plates slide toward one direction or another and, inevitably, their ridges move off the hot spots that found them.

As the plates perpetually shift, they expand, recede and slide across the face of the earth in utter disregard of the hot spots drilling up through them. "The subduction zones, the numbers of subduction zones, the mid-ocean ridges, their numbers, their distribution, their orientation with respect to each other, all of those things are chaotic," says David Stevenson at Caltech. "They're being continuously reorganized. If you got into a time machine and went 200 million years into the future, you wouldn't recognize the earth."

An instance of this change is the Mid-Atlantic Ridge. It has moved off the line of hot spots that gave rise to it. For the last thirty million years it has been drifting westward. Similarly, the fiery spout of Tristan da Cunha, which once straddled the ridge, is no longer burning a line of volcanoes into the South American Plate. Like most of the other island hot spots, it lies some five hundred miles to the east, solidly beneath the African Plate.

On the other side of the globe, the push-me–pull-you nature of tectonic motion has done away with a number of plates since Pangaean times. Researcher Tanya Atwater of the University of California at Santa Barbara pieced together this case of the missing plates years ago. Approaching the problem like a dedicated detective, she was so eager to solve it that she neglected the Ph.D. thesis she was supposed to be finishing. Such was the excitement in the field as the new theory of plate tectonics was taking shape.

As Atwater remembers it, she was sitting with a handful of geologists one evening, watching as they animatedly sketched the outlines of the plates on cocktail napkins. Using scissors, they demonstrated how the crustal building blocks moved around. She was completely taken with the simple and elegant geometry. Soon she became fascinated by the idiosyncrasies of plate movements along the western edge of the Americas and proceeded to reconstruct the plates of the ancient Pacific, most of which were lost into a trench that once had lined the California coast.

"The most important thing is to realize that in the past there were lots of oceanic plates, not just one big Pacific," she says. Her reconstructions show that the Pacific Plate has grown enormously in the last 170 million years. Formerly a tenth its present size, the Pacific's rapid expansion has crowded out at least three other oceanic plates that once surrounded it. Along its northwestern border, probably next to the Asian continent, lay the Izanagi Plate, and along its southern border lay the Phoenix Plate. For a brief time a rapidly moving plate also existed to the north, but it has since disappeared into the Aleutian Trench. Known as the Kula Plate, it was named after a word from the Athapaskan Indians of Alaska meaning "all gone."

How did Atwater know these plates existed if they have all disappeared? As luck would have it, the missing plates left behind enough clues so that she could piece together where the plates had once been. Each plate shared a spreading ridge with the Pacific Plate, a ridge that has long since disappeared down the ocean trench. But Atwater knew where the plates were because the magnetic stripes on the ocean floor get younger and younger as they approach the now defunct Izanagi, Phoenix and Kula Ridges. She could therefore infer that these ridge crests, as well as the plates on their far sides, once bordered the Pacific. All were eventually shoved aside as the East Pacific Rise spewed out new crust at a much faster rate.

This hard-driving ridge was also laying ground on its eastern side, building up the huge Farallon Plate that once rested beneath most of the Pacific Ocean. But faster than the ridge could generate new crust, the Farallon was disappearing into a trench that, until about 25 million years ago, ran all along the western coast of North America. Now only a tiny fragment of the plate and its ridge remain off the shores of Washington.

Atwater says that the prolific East Pacific Rise, which repaved that ocean's floor, may also disappear one day. Between the ridge and a trench along the Peru-Chile coast lies the Nazca Plate. Faster than new crust can be added to the plate, old crust is being consumed by the trench. As the Nazca slowly loses ground, the East Pacific Rise is inching toward its demise.

This is the way that most of the world's surface shifts. The tectonic plates are no more permanent than the sovereign countries of the world. Their boundaries shift with the quick flow of the shallow mantle. And as it is with frictions between nations, every twitch and scrape along the plate borders holds the potential for a major upheaval.

The Language of the Rock

Even with advances in the field, however, crucial pieces of information are missing, especially in these studies of the surface. One of the main reasons is that geophysicists have been strangers to the field of geology for such a long time. In many researchers' minds, and for obvious reasons, deep-earth science leaves off where surface science begins—in other words, where geology begins. For decades there has long been a lack of communication, some even say a disdain for each other, between the two fields.

As geologist Christopher Scotese, at the University of Texas, explains it, geophysicists have always had the upper hand in terms of making models and amassing numbers. Theirs was the dominant field, so there has been resentment as a result of that. The main reason, however, is that each group has defined itself and its work in a certain way and each has continued along that path. Geologists and geophysicists rarely go to the same meetings or collaborate on the same problems.

Geologists have all grown up believing that the rocks tell a story. The layers upon layers of stone are all laid down in a specific sequence, and those who know how to read that sequence can read the story of a region—when uplift took place, when volcanoes loosed their contents. Geophysicists, on the other hand, have become so specialized in their subfield that few of them know how to read geologists' maps or understand the subtleties of strata. This parochialism leads many geologists to believe that, without the hard evidence, geophysicists are missing a big piece of the earth's story.

Scotese is one of the researchers bringing an end to this ignorance. He has created a series of maps of the earth's surface going 545 million years back in time. These "paleogeographic" maps take as their springboard all the geologic data gathered over the decades. They attempt to show the evolution of any particular acre of land anywhere in the world. His maps reveal where the land, the sea, the mountains, the subduction zones and rifts were throughout the eons and where shallow seas once covered the continents. Thus Scotese has assembled all the pieces of information in a way that brings out the geophysical underpinnings of a region.

Scotese created these maps for his own geological interest, not in any attempt to bridge the gap between geology and geophysics, but his colleagues in geophysics have seized upon the wealth of data the maps provide. With this kind of information they can begin to coordinate what has happened on top of the earth with what's going on inside.

Combined with simple models of mantle convection, the maps have already thrown new light on sinking slabs. Geophysicists believe they see these slabs tugging on the skirts of the continents as they subduct, letting the ocean waters push far inland along these boundaries. That recent insight came from pared-down models of both surface and mantle. Many researchers think that the next step will be to put the interior and exterior together in all their complexity. For the moment, though, this is still beyond the limits of computer power.

Collaborations such as this take advantage of the geophysicist's ability to model the dynamic flux within the planet and the geologist's record of what has actually gone on in the past. Here is where researchers believe they will find a complete portrayal of the planet's character. For the history of the earth is in many respects like the history of our own lives. The true meaning of anyone's life can be seen only by looking at the entire arc of one's existence. Similarly, changes take place so slowly on earth that the most interesting aspects of the planet's activity can be seen only through the binoculars of time.

Some of those changes are akin to the gray hairs and weakened gaits that people develop as they age. Over the eons, the earth is gradually cooling. As it loses the heat that fuels its steady churning, the plate motions on the surface are slowly winding down. This progressive, predictable effect stands in the background of the earth's life.

But superimposed upon this general decrease is a more remarkable series of events. Pangaea's crack-up was a single episode: a period in earth history of rapid change and rapid plate motion that stood between periods of relative quiet. Yet this cataclysmic event now appears to be part of a much broader cycle—one that has held sway over the planet's surface throughout its history.

10

THE MUSIC OF THE SPHERE

Many geophysicists now believe that the Pangaean supercontinent was not a onetime phenomenon, that at least once and probably several times in the past similar supercontinents have amassed and then subsequently broken apart. In and out, throughout the eons, the ancient continents must have drifted. This global interplay of the landmasses has come to be known as the Supercontinent Cycle.

It is an idea that geophysicist Don Anderson first began pondering in 1982. He realized that since heat conducts only half as quickly through continental crust as through oceanic crust, the former acts like an insulating blanket. A mass as large as a supercontinent would cause the mantle to grow hotter, which would affect circulation. Says Anderson, "Eventually I believe the mantle gets so hot that it will form currents that are hot enough to break up the supercontinent."

To Anderson's eye, the cluster of hot spots and the bulge

in the earth exactly beneath the former site of Pangaea not only broke up the supercontinent but were formed precisely because it was preventing heat from getting out. When tomographic maps later showed a patch of warm rock beneath that spot, it further confirmed his belief. The giant Pangaean landmass sowed the seeds of its own destruction.

Based on this idea, Anderson suggested an even more intriguing notion. He saw Pangaea's breakup as an episode in a broader cycle in which the earth's surface has alternated between supercontinents and fragments, all as a result of these landmasses insulating the earth's interior.

In 1988, geophysicist Michael Gurnis at the University of Michigan worked out the mechanics of Anderson's Supercontinent Cycle. His simple simulation of the cycle represented the world with a rectangle. Inside this box, the hot, red upwellings of the mantle moved past cold, blue downwellings. On the surface of the box floated horizontal bars: the schematic continents. Although unrealistic by earth standards, Gurnis's two-dimensional model was the first in which the continents were able to affect the flow in the mantle as well as the other way around. In it Anderson's predictions are vividly confirmed.

In this view, hot, insulated mantle wells up beneath a supercontinent, causing it to balloon upward. Then suddenly the landmass tears apart. It's as if, as one group of researchers described it, the world were a coffee percolator. The amount of heat it emits doesn't change. But because that heat gets trapped by the continents, it comes out in dramatic bursts. After rifting, the continental fragments in the model quickly slide toward cold patches of the mantle. Eventually they reaggregate over a cold spot, forming a new supercontinent. When this landmass begins to warm up that section of the mantle, the cycle begins anew. The model continents repeatedly come together and break apart roughly every 500 million years.

"I think what that says," comments Arthur Lerner-Lam at the Lamont-Doherty Earth Observatory, "is that the disposition of the continents on the surface of the earth . . . has a lot to do with modifying the planet's whole convection system, and that we're not seeing a system that's merely passive on the surface but something that's intimately involved with a deeper pattern."

Once again the crust isn't merely reacting to interior pressures. But this time, instead of the plates impinging on the interior, it is the continents—which seems to introduce yet another layer of complexity to the earth machine. Or does it?

The latest flushing models of mantle convection have shed new light on what may keep the supercontinents dancing. Larry Solheim, who, with his colleague Richard Peltier of the University of Toronto, built one of the earliest two-dimensional flushing models, thinks this peculiar aspect of convection is linked to the Supercontinent Cycle. "The avalanche can bring in surface material from up to about 10,000 kilometers [6,000 miles] away," he explains. "So you'd have this local sinkhole, this avalanche of fluid across the boundary, which would act like a big vacuum pulling material from the surface into this central location. All the continents would come and crash together above this sinking fluid." Solheim says that heat would then build up under the supercontinent, causing it to break and then spread apart until the next avalanche drew the pieces back together.

The flushing models have thrown a curve to geophysicists trying to work out how this cycle has come about. Is it purely a result of the insulating effects of supercontinents, or is it a tug-of-war between flushes pulling the continents together and heat breaking them apart? Whatever the answer, this insight into how the earth's surface affects its interior has provided geophysicists with an interesting new way to look at how the earth's surface has behaved over the last four and a half billion years.

Supercontinents Through the Eons

Some 500 million years before Pangaea was formed, all the land in the world seems to have assembled into another gargantuan island. This long-lost supercontinent has been christened "Rodinia" after the Russian verb *rodit,* "to beget." Although little is certain about the arrangement of Rodinia's landmasses, geophysicists do know the pieces it broke into, because these continents came back together to form Pangaea. The proto-Pangaean continents left a record of their existence in the mountain belts built by their collisions. Indeed, most of the mountain ranges on earth today owe their origins to this barrage of continental fender benders.

The Caledonian Mountain Belt, which extends from Scandinavia down through the British Isles, traces back to this era. Some 400 million years ago, a great landmass that included North America, Greenland and part of the British Isles collided with Baltica, a continent containing most of northern Europe and Russia. This huge assemblage then collided with another great continent known as Gondwanaland, which included South America, Africa, India, Antarctica and Australia. The crunch between these two massive continents raised the Appalachians, beginning around 300 million years ago. With this collision Pangaea was almost complete, except for the three blocks of land that converged onto Baltica at the eleventh hour. Siberia, Kazakhstan and China collided to raise today's

Urals of eastern Europe and the Altai and Tien Shan Mountain Belts in northwest China.

So although researchers don't have the tracks on the ocean floor to retrace the movements of the plates that carried these continents, they do have the record of breakup and assembly that is embedded within them. The history of the continents has moved along a repeating loop of rifting, mountain building and then rifting again. At least back to 570 million years ago, it lives up to the predictions of the Supercontinent Cycle.

Prior to this lies the yawning stretch of epochs known as the Precambrian. This extends all the way back to the formation of the earth about 4.5 billion years ago. Rocks that have endured from the Precambrian are highly deformed; they've been squished, baked, sliced and stretched for eons. Despite the obscuring veil of time, many researchers can make out the traces of the Supercontinent Cycle in the Precambrian. Periods of mountain building tend to clump into six distinct clusters starting with the bout of continental collisions that formed Rodinia, and recurring roughly every 500 million years back in time.

By far the most detailed picture of the continents aggregating and dispersing in this ancient era is due to the efforts of Paul Hoffman, a geologist at the Geological Survey of Canada in Ottawa. Hoffman possesses an encyclopedic knowledge of Precambrian continental geology. Throughout the 1980s he sifted through the complex early history of North America's development.

Aiding his efforts was an improved method for dating rock known as uranium-lead isotope analysis. This new technique has allowed geologists to determine the age of 2-billion-year-old rocks to within a few million years—a vast improvement over previous methods. Hoffman used a huge collection of rock ages that both he and other researchers had gathered to compose a sweeping synthesis of the North American continent's genesis. The landmass appears to have come together in a series of stages which Hoffman, a classical music buff, likens to a symphony in four parts.

The first movement is an allegro in which seven microcontinents were briskly fused into the heart of ancient North America, a landmass known as Laurentia. As the microcontinents collided, they piled up the sedimentary rock along their shores into mountain belts. Those peaks have since eroded, and all that is left are the sutures between the microcontinents, like the mortar between two bricks. These range in age from 1.96 to 1.81 billion years old. Thus the bulk of Laurentia came together in the relatively short time span of 150 million years. That it took so little time to assemble, Hoffman wrote in his original 1988 paper, "is the single most startling conclusion to emerge from this synthesis." Instead of microcontinents bumping up against each other in a sort of Brownian motion, they were all swept together in a sudden whoosh, like motes of dust scurrying toward a vacuum-cleaner nozzle.

Continuing the symphony, Hoffman says, "the second movement combines the dominant themes of the movements before and after." Between 1.8 and 1.6 billion years ago, a huge chunk of land struck and stuck to the southern border of Laurentia. It annexed the states from southern California to Michigan as well as part of southern Sweden. At the same time, a new phase of activity began. Hot magma bubbled up through the older parts of Laurentia west of Hudson Bay, signaling the start of continental rifting.

For the next 300 million years, the outpouring of magma built toward a crescendo. Phenomenal floods covered thousands of square miles of Labrador and spread like a vast sheet across the central and southwestern United States. That layer of igneous rock now underlies the soil throughout most of America.

During this third movement, an adagio, the land also developed stretch marks. Wall-like features known as dikes (which form when magma pushes into a crack and then solidifies) cut across thousands of miles of the Canadian Shield. They are a sign, Hoffman believes, of the hot pancake of magma that was warping the crust from underneath.

But Hoffman's analysis shows that the outpouring of lava and the stretch marks were the greatest event of Laurentia's rifting. The continent never broke into the tiny fragments that marked its beginning. That, he explains, is probably because Laurentia was the hub of this Precambrian supercontinent. Like Africa today, it may have stalled above the mantle upwelling while the other continents dispersed. Then, before rifting was complete, Laurentia slammed into a large, unidentified continent to the southeast. Three thousand miles of crumpled crust, from southern Mexico to southern Sweden, heralded this symphony's final allegro. A new supercontinental aggregation had begun.

Hoffman's results don't yet describe what was happening on all of the other continents at the time of Laurentia's symphony. "Ultimately," says Hoffman, "I'm intending to try to make a synthesis similar to what I did for North America but for the global Precambrian." Thus he and other researchers are gradually fleshing out the Supercontinent Cycle. To convince themselves that this cycle is real, researchers will have to amass the geological evidence for more than the two recent episodes of Pangaea and Rodinia.

The value of finding the Supercontinent Cycle in the Precambrian goes beyond the reconstruction of our lands in that distant era, for geophysicists believe there was more to this cycle than a treadmill of rifting and mountain building. The continuous wandering of the continents also produced a cascade of side effects.

The Deluges

"Perhaps the most vivid geological observation we have is that the flat areas of the continents were actually completely covered by great inland seas," says Michael Gurnis. Time and time again the global oceans overflowed their basins and spilled out onto all of the continents. And geophysicists believe a connection exists between this phenomenon of global flooding and the Supercontinent Cycle.

The links first began to take shape during the early 1980s, in the minds of three researchers who were living and working in rather close quarters. They are Tom Worsley and Damian Nance of Ohio University and Judith Moody, president of J. B. Moody and Associates in Columbus. At that time Worsley, who is married to Moody, had also struck up a friendship with Nance. Worsley was living and working with Nance on weekdays and then commuting back to Columbus, where he and Moody lived on weekends. Sharing their lives in this way motivated the three researchers to look for the connections between their various fields.

"What happened," says Worsley, "was that we kept looking for common ground as an excuse to work together. I really like things having to do with sea level, being a paleontologist. Moody was essentially the Julia Child of basalts. She used to cook basalts for a living and see whether she could alter them, and Nance was interested in the tectonics of an entire continent. Suddenly one day I sat down and began some back-of-the-envelope thoughts, and probably in one afternoon I called them both in and said let's look at this."

Out of this mix of personal and professional motivations came the researchers' realization that they could explain the sea-level changes if they focused on what was happening during a supercontinental breakup. In fact, with this first step, they were retracing the steps of many other scientists. As researchers have known almost since the inception of the plate-tectonics theory, the rifting of continents can explain most of the changes in global sea level. That's because these rifts determine how much water the global ocean can hold.

The bottom of every ocean is shaped like a garden-party canopy, with a bump in the middle and a smooth slope to the shallow basins on either side. That bump is the mid-ocean ridge, which is forced up by the hot magma beneath its crest. The high ridge displaces ocean water. Therefore, the newer an ocean, the shallower it will be, because more of it is made of this hot ridge crust. When supercontinents break apart, they have an enormous impact on the volume of water the global oceans can hold. The effect on sea level is on the scale of hundreds of yards, about three times larger than the changes wrought by gravity.

With this in mind, the Ohio researchers calculated when sea level should be high and low and matched these findings against a geological record known as the Vail curve. This curve is an index of global sea levels compiled from oil exploration data by Exxon scientist P. R. Vail. It extends back 570 million years, as far as geologists can go on solid data. The Ohio researchers' predictions were a close fit with the Vail curve but, as others had seen before, not quite perfect. Not quite, that is, until the researchers seized upon another idea.

Not only do supercontinents break up and reassemble because of the heat welling up beneath them, but they also bob up and down—again, in a range of several hundred yards. To the aging and birthing of the seafloor the researchers added the effects of this supercontinental bobbing. Together the two motions should, they believed, reflect sea level throughout the ages.

Indeed the motions did. Dramatically high sea levels accompanied the dispersal of both Rodinia and Pangaea. Both times the global sea was some three hundred yards above its present height, submerging about a third of the land we see today. Between breakups the continents stood high and dry.

These sweeping changes in sea level paralleled the Ohio group's model of rifting and bobbing landmasses. To be sure, their calculations were only global averages, and sea level varied from ocean to ocean. Local effects, like glacial melting, must also be taken into account before the Ohio model can account for all the coastline changes that researchers see. Nevertheless, if geophysicists can believe that the earth's surface is continuously rippling with the waltz of the continents, Worsley and his colleagues have shown that the global ocean is stepping in time.

All that water has done more to shape and color the landscape than perhaps any other activity on earth. The great inland seas deposited layer after layer of sedimentary rock onto the continents—flatirons of tawny sandstone and gray-green shale. As the midlands have been buoyed upward by the earth's heat, rivers and streams have carved through the stone stacks, producing the magnificent curlicues, arches and canyons of the American Southwest. The spectacular Mesa Verde, where prehistoric Indians carved cliffside aeries into a thick layer of sandstone, originated in the inland seas of the Late Cretaceous period after Pangaea had broken up.

At the same time, thick forests surrounded North America's inland sea along an eastern shore that moved back and forth across the Midwestern states. Whenever the waterline advanced, the vegetation was swamped. When the water receded, the trees sprouted anew. Many layers of decaying organic matter built up in Pennsylvania during the Cretaceous, and now it is all coal. Great beds up to 250 feet thick underlie the state. It is not too much of a stretch to say that the waxing and waning seas of this period fueled our Industrial Age.

Worsley and his colleagues have extrapolated these effects of the Supercontinental Cycle back into the Precambrian. In their opinion, the world's landmasses have been bobbing up and down like floats on a fishing line while ocean water has periodically flooded them for at least 2 billion years and perhaps longer. Geophysicists' ability to fix sea levels during the earth's early history is spotty at best. But just as researchers like Hoffman are trying to pin down the cycles of rifting and mountain building during the Precambrian, Worsley's group and others are trying to flesh out the supercontinents' effects on the sea.

Into the Skies?

If these researchers are right, the ramifications of this continental shifting may also extend beyond the oceans and solid ground. Cycling in synchrony with the supercontinents and sea levels is the global climate. As Alfred Fischer, one of the elder statesmen of geology at the University of Southern California, pointed out several years ago, whenever sea levels dropped in the last 570 million years, great sheets of ice tended to form on the continents. Such glaciers covered most of the Pangaean supercontinent from 250 to 300 million years ago, and the ice age that blanketed the Rodinian supercontinent was one of the worst ever. Conversely, whenever the continents were flooded by oceans, the climate warmed. During the Cretaceous period, after Pangaea's breakup, the earth experienced a hot and humid climate that ushered in a golden age of biological productivity.

Scientists have traditionally attributed such broad climatic effects to changes in the amount of carbon dioxide in the atmosphere. A greenhouse gas, carbon dioxide is able to keep the sun's radiant heat from escaping back through the earth's atmosphere. And since climate changes seem to go hand in hand with sea-level changes, Worsley and his colleagues have proposed that the Supercontinent Cycle may be the ultimate source of the shifts in carbon dioxide.

The mechanism they propose is relatively straightforward. Over time periods of 500,000 years or more, carbon dioxide in the atmosphere rises and falls with the geologic cycle. Volcanic emissions are the main suppliers of carbon dioxide to the climate system, while continental weathering is the main activity depleting it. In this latter process, carbon dioxide dissolved in rainwater combines with limestone to form bicarbonate and calcium, which are used by ocean creatures to build their shells, effectively removing the greenhouse gas from the system. The more the continents are exposed to weathering, the more carbon dioxide is drawn down from the atmosphere. Worsley and his coworkers believe that the Supercontinent Cycle may dictate the global climate simply by changing sea levels, and thus the area of the continents exposed, as well as by turning volcanic activity up and down when landmasses fragment.

It is difficult enough to know what our atmosphere is doing from day to day, and climate modelers are loath to say that such a theory on what the weather was like hundreds of millions of years ago is anything more than speculation. "One can make quite a nice *Just So* story," says climatologist James

Kasting at Pennsylvania State University, "but whether or not it can convince our colleagues is another question." Researchers are still not sure of the effects of changes in the earth's orbit on temperature, of global warming and glacial melting on sea level, of mountain building on the rock exposed to weathering, or of the feedbacks between all of these factors.

Moreover, researchers are not sure how far back into the Precambrian any of these influences applies. Four billion years ago, the heat flowing out of the earth was as much as four times higher than it is today. Yet the solar radiation from the sun, then a faint, young star, was approximately 30 percent less. Through this mixture of internal and external heat plus early greenhouse gases, temperatures are thought to have been uncomfortably hot by human standards. Yet the first global ice age may have come as early as 2.5 billion years ago.

While the climatic implications are still somewhat fuzzy, Don Anderson's global Supercontinent Cycle and the Ohio team's explanation for continental flooding are far more concrete in most researchers' minds. They offer one of the most intriguing views of the earth to emerge in recent years. This image of the never ending machinations of the planet also harks back to an idea that earth scientists have had for generations. In the eighteenth century the Scottish geologist James Hutton published a landmark treatise, *Theory of the Earth,* in which he set forth the principle of uniformitarianism: the present is the key to the past. He proposed the same processes of erosion, weathering, uplift and sedimentation that are sculpting the land today have been at work throughout the eons. They are all that is needed, he said, to explain how the landscape around us came to be.

In Hutton's day, uniformitarianism was a refutation of theories that called upon catastrophic events like the biblical flood to explain the way the world looked. But even as these deus ex machina theories fell by the wayside, Hutton's ideas continued to hold sway in the minds of earth scientists. When the plate-tectonics revolution overturned their picture of the world, researchers had to make room in their definition of uniformitarianism for elements like continental rifting and the disappearance of the surface crust down ocean trenches. But still the image of a continuum persisted.

Now that notion is being expanded again to include deeper processes. The magnetic field reverses polarity. Plumes rise and erupt on the surface. The crust and mantle churn together on both the yearly scale of incremental plate motions and the epochs of supercontinental cycling. Adapting uniformitarianism to fit the 1990s, geophysicists even make room in their modern theories for occasional catastrophic events like mantle avalanches or, as we shall soon see, asteroid impacts. Yet these have not brought the principle of uniformitarianism to its knees. The more things change, the more they stay the same.

The image of the Supercontinent Cycle adds yet another twist to this age-old theory. Rather than just evolving in a gradual, uniform manner, the earth may actually be caught up in a repeating cycle. Yet the same processes at work today were also present in the past. Supercontinents of old are just as important as weathering and sedimentation in helping researchers explain the way the modern landscape arose.

Though many of the ideas upon which the Supercontinent Cycle is based are still being debated, this image brings together many of the pieces of the dynamic-earth concept. It is thus a first step toward a modern-day theory of the whole planet and a good picture to hold in mind as we move on to a more specific look at the ways that activity inside the planet impinges on us at the surface.

The motions of the earth are woven into our daily lives in myriad ways. Earthquakes rattle certain spots within the crust, and volcanoes explode with demonic violence. These are the geologic events that affect us all most personally. They demand our attention with the same immediacy as the everyday crises in our lives. For that reason, earth scientists throughout the centuries have devoted most of their time to these palpable events, and modern researchers are no exception. But with their newfound vision of the earth, researchers are more sensitive to the interconnectedness of cataclysms on the surface and the dynamic interior.

The Supercontinent Cycle alone has left the continental crust riddled with the scars of former rifts and mergers. Geophysicists have examined these continental scars to tell the story of cataclysms that struck the earth long ago. But the break that was once caused by a balloon of hot rock rising from the surface of the core turns out to be of more than historical interest. It may be just the flaw that will set our modern landscape to trembling.

11

TIME BOMBS

Before Big Muddy went on its rampage of 1993, swallowing towns and the checkerboard of farm plots all along its banks, the Mississippi River received media attention for a different reason. Where it crosses through the bootheel of Missouri is the site of the Reelfoot Rift, a seismic fault line smack in the middle of the North American continent. That was the rift that grabbed headlines late in 1990, as a result of a dire forecast. Iben Browning, a New Mexico climatologist, proposed that the entire region, including the tiny town of New Madrid, Missouri, would be shaken by a disastrous earthquake.

Browning based his prediction on the fact that the sun, the earth and the moon would align during the final month of 1990, raising tides in the earth's crust that are similar to ocean tides. Browning claimed that the strain from these tides could be enough to trigger an earthquake in the Reelfoot region. And he gave the chances as fifty-fifty that a large quake would strike New Madrid on December 3, 1990, or within forty-eight hours of that day.

Geophysicists were doubtful. They have repeatedly looked for a connection between crustal tides and earthquakes over the past few decades, but to no avail. They let the citizens of New Madrid know that a major earthquake was no more likely to strike on December third than on any other date. But the people who lived in this town weren't taking any chances.

Schools closed. Residents planned vacations out of town. The governor of Missouri, John Ashcroft, and one of its U.S. senators, Christopher "Kit" Bond, scheduled visits to New Madrid to coincide with the expected quake. Even the Missouri State Penitentiary, two hundred miles away, took fingerprints and toeprints from its inmates in case some prisoners turned up missing in the earthquake's aftermath.

Not surprisingly, Browning's prediction didn't pan out. The third of December came and went without a ripple. The scores of journalists who had descended on New Madrid for the fateful day ended up reporting on one another instead. And with all the tourists in town, the handful of taverns and grills along Main Street did a booming business.

Yet that kind of hysteria didn't rise from mere fancy. Behind it lay a tumultuous precedent—one of the most disastrous incidents to beset the face of the earth. Early in the last century, the three most powerful earthquakes ever to strike the stable heart of a continent rattled the Reelfoot Rift. They came at a time when the Mississippi River boasted a steady traffic of steamers and flatboats carrying supplies to western settlements. New Madrid was a small river outpost. It consisted of a smattering of stores, a tavern, a post office and a few hundred residents, mostly farmers, on widely spaced plots.

The first earthquake struck this tiny Midwestern community on December 16, 1811, at a little after two A.M. The shock was a magnitude 8, and from then on the ground around New Madrid rumbled continuously. Most of the town's few hundred settlers moved into tents and other temporary shelters after the first earthquake, thus saving themselves from the real damage that was to come. On January 23, and again on February 7, two more magnitude 8 quakes shook the earth.

One by one these mighty quakes rocked the central Mississippi Valley and then spread throughout the nation. Church bells rang in Boston, a thousand miles away, and crockery rattled all the way from Middlebury, Vermont, to Savannah, Georgia. Closer to the source, the ground showed more spectacular signs of the earth's foment. As James Lal Penick has documented in his detailed account *The New Madrid Earthquakes,* observers recalled seeing great geysers of sand, water and carbonized wood spewing from the ground. The intense shaking had turned wet sediments into a pressurized liquid that exploded toward the surface. These "sand blows" erupted with a loud bang and then a roaring whistle like the shrill of a desert sandstorm.

Along the riverbed, the sand blows created funnel-shaped holes that whipped the water near them into deadly whirlpools. These "sucks" compounded the chaos that overtook the Mississippi. As the land undulated, waterfalls appeared and disappeared. For a few hours after the February shock, witnesses even saw the mighty river run backward. And as chunks of the Mississippi bank dropped suddenly into the river, they raised mountains of water that capsized dozens of boats and claimed an unrecorded number of lives.

Eventually the river swallowed New Madrid itself. The last and greatest of the three quakes, now estimated at magnitude 8.3, leveled the town's vacant houses and caused the bank beneath New Madrid to sink a dozen feet to half its original height. After that, the town continued to sink until spring floods finally washed it away. Most of New Madrid's inhabitants survived and rebuilt their homes. But in nearly two centuries since this catastrophe, the region has repeatedly endured lesser though still damaging earthquakes, and more are expected. The unrelenting question is, Why?

Geophysicists cannot yet predict when any earthquake will occur, and no one can say whether that capability is years or decades down the road. But for the vast majority of tremors, researchers at least know where on the globe to expect them. Earthquakes line the borders of the tectonic plates and are symptoms of the perpetual motion inside our planet.

The theory of plate tectonics, though, has no bearing on the New Madrid quakes. They occurred nowhere near a plate boundary. Instead they were right in the middle of the giant North American Plate. Although this plate moves along with the earth's internal motions, the stress of that movement is supposed to be concentrated along its borders. The same is true for all the plates. According to the tectonic theory, they should behave like rigid slabs; and most of the time they do.

What's more, the New Madrid quakes were not the only examples of earthquakes erupting well within the boundaries of their plates. On August 31, 1886, a magnitude 7.6 quake rattled Charleston, South Carolina, leveling most of the city. This was the second-largest mid-plate quake recorded in North America, after those in New Madrid. Most recently, a magnitude 5 earthquake struck at the border of England and Wales on April 2, 1990.

Only a decade ago, most researchers viewed these mid-plate earthquakes as isolated or random incidents. But a new geophysics, of dirt, has emerged in recent years, and with it researchers can date prehistoric mid-plate quakes. Paleoseismology, as the new field is called, is science by backhoe. Researchers dig deep, straight-walled ditches and search up and down the wall of earth for signs of shaking. The sand blows of the New Madrid earthquakes, for example, left conspicuous blast cones in the region's dirt. These cones also collected organic sediments like tree fragments, which can be carbon-dated. Using the techniques of paleoseismology, researchers have now amassed evidence that earthquakes have struck some mid-plate regions repeatedly in the past.

A total of six large earthquakes, including the one in 1886, have racked Charleston, South Carolina, in the last two thousand to five thousand years. In the Wabash Valley, an area along the Indiana-Illinois border that has had five earthquakes of magnitude 5 or greater in the last two hundred years, geophysicists have found signs of an even larger earthquake that struck before settlers arrived in the region.

Compared to the number of earthquakes that have rattled California's San Andreas Fault, these mid-plate earthquakes may seem few and far between. Paleoseismology has nevertheless shown them to be afflicting certain areas with some regularity—and with disastrous consequences. Because these earthquakes strike in the strong, unfractured rock of plate interiors, their tremors travel farther than those from border quakes before petering out. The New Madrid quakes, remember, shook the nation's eastern seaboard. In comparison, the tremors from a magnitude 8 earthquake on the Alaskan Plate boundary on Good Friday, 1964, traveled only a few hundred miles before they subsided. The possibility of such a mid-plate quake thus carries a much higher risk than one on a plate boundary. That thought has sustained geophysicists in their efforts to figure out what is special about the parts of the continent where these mid-plate quakes strike.

Recently Arch Johnston, director of the Center for Earthquake Research and Information at Memphis State University, and his colleagues discovered a major clue. They studied the tectonic settings of nearly nine hundred earthquakes that took place in continental interiors and found that the majority of these earthquakes came from crust that had once been stretched. That description applies to either the continental crust lining an ocean or crust that started to split, as eastern Africa is doing now, but for some reason didn't complete the break. "If you lump these together as extended crust," says Johnston, "they make up about one-quarter of all stable continental crust, but they have about two-thirds of the earthquakes greater than magnitude six and a hundred percent of those greater than magnitude seven." Johnston and his colleagues believe that almost all mid-plate quakes have taken place in such weakened areas.

The Reelfoot Rift, a thinned and fractured band of crust, was first deformed some 550 million years ago, when the Rodinian supercontinent was breaking up. At that time, a hot dome of rock was pushing beneath what is now the central Mississippi Valley, causing the continental crust to balloon upward and thin. Slowly, a rift that ran from southern Texas to eastern Oklahoma was making its way up through Arkansas and Missouri, fracturing the continental crust. This rifting ceased before the crust broke completely, but the episode left the Reelfoot area weak compared to its surroundings. Eventually the stretched crust gave rise to the New Madrid earthquakes, and it will undoubtedly spawn others.

Johnston's investigations into the continental crust have added a new component to the theory of plate tectonics. Like a cracked china cup, the continental crust is still fragile where it has been damaged in the past. Wherever hot rock inside the earth has risen up to break or merely to stretch a strip of continental crust, it has left a zone of weakness that can come back to haunt us. And that scar has nothing to do with the present borders of the plates. Thus as researchers have begun to put together the earth's inner turmoil and its effects on the surface, they have also faced an added threat. It appears that geophysicists have to worry not only about the risk of earthquakes from present plate interactions, but from past interactions as well.

Potential disaster now lurks in the areas where researchers least suspect it. "You can't paint every old rift red on the map, because there are too many of them," says Mark Zoback at Stanford University. "And there are too many of them that aren't active. There are huge, old rifts all through the central U.S. which don't seem to produce any major earthquakes now." So the real mystery of mid-plate quakes is why we don't see them everywhere we turn.

The selective aspect of where these mid-plate quakes strike has led Zoback to believe that another element besides old rifts has to be involved. The forces driving the plates and the stretched zones in the crust must somehow conspire with a third factor to trigger these mid-plate quakes.

One clue is the odd fact that the active rifts haven't produced much deformation. Slipping, shuddering faults should stir up the terrain in much the same way that movement along the San Andreas Fault has built up mountains in California. Even if mid-plate earthquakes are far less frequent than those on plate boundaries, if they have continued for any length of time they should have shoved up some hills. Yet geophysicists don't see these telltale signs. "For some reason these areas turn on and off over geologic time," says Zoback. "So when we look at these places like New Madrid and Charleston, they're turned on now and they've been turned on for a few thousands of years, and we can see a few of these earthquakes occurring with these more or less thousand-year recurrence times. But they haven't been turned on forever, and they're going to turn off at some point because we just don't see the cumulative deformation that one would expect to see had those areas been persistently active over the geologic record."

The flip side to this notion is rather grim: at any given time or place, another one of these faults might suddenly turn on. Researchers now find themselves waiting for the Big One somewhere, perhaps anywhere, in the nation's heartland. That leaves them queasy about their role in protecting society from these disasters.

"There is the unpleasant feeling," says Klaus Jacob at Lamont-Doherty, "that each of these major earthquakes seems to crop up in an area where the historic record has not indicated that they would. And that's rather disconcerting, because if you have either structures such as a nuclear power plant or a major city somewhere that hasn't had an earthquake that is of consequence in the historic record—let's say three hundred years at most in parts of this country—that does not necessarily mean you can conclude that it wouldn't have one in the next three hundred years."

For communities along the San Andreas Fault, where there has long been a clear seismic threat, earthquake-preparedness measures are largely in effect. Building codes have been adjusted, and the public has been educated about what to do when an earthquake strikes. Where deaths have occurred, such as those in the 1994 Northridge quake, they have largely been the consequences of not following those building codes properly.

Most cities have taken no such measures or have taken them so recently that few buildings conform. A case in point: Memphis, the largest city near New Madrid, adopted earthquake building standards in 1991, but they apply only to future buildings. It will take Memphis a long time to outgrow its existing structures, though at least its citizens are aware of the threat. Pressing upon the rest of us is the image of all those dormant scars in the crust potentially surging to life.

The only preventive measure researchers can take is to

concentrate on what is triggering quakes on the rifts that *are* turned on. While tentative theories abound, many researchers are focusing on how the surface plates are locked in step with the rock currents of the interior. Is there something about the lower crust, the soft glue-like rock that binds the plates to the moving mantle, that is putting added stress on some parts of the system—stress that gets released at the weak spots in the upper crust? Zoback, for example, believes the key may be that the lower crust beneath these mid-plate quakes is able to deform more easily.

He says that one factor that could cause this added deformability in the lower crust is a warm patch of rock beneath the region. The heat would increase the malleability of the rock above it. Zoback and a graduate student, Lanbo Liu, recently reported that surface measurements showing slightly more heat flowing out of the ground around New Madrid than the surrounding crust lend support to this idea. Their theory offers up the interesting possibility that these shallow quakes have their source in the deeper earth.

Yet seismic tomographers, blanketing the region with thousands of crisscrossing surface waves, have found not a warm patch but a cold spike of rock directly beneath the New Madrid region. Dave Yuen at the University of Minnesota thinks the spike may be pulling down on the crust around New Madrid and thus causing the mid-plate earthquakes, though he has yet to crunch through all the data to explain how this would work. In addition, tomographers are increasingly warning their colleagues that near the earth's surface the speed of seismic waves is less likely to be a gauge of the rock's temperature. Yuen's spike may turn out to be not cold but simply rock with a different chemistry that makes seismic waves travel through it quickly.

For now, researchers admit that they are still grasping at

straws. The true cause of mid-plate quakes remains one of those unsolved puzzles about the workings of the earth machine. But what this example shows is how many different aspects of the earth go into creating a single event, even one as seemingly rare as an intraplate earthquake. The cause of the quake appears to draw on forces from both the surface and the depths, from activities of the present as well as the distant past. Somehow the interplay of a whole host of factors can add up to push the stable crust out of balance.

This commingling of forces and events is true for a wealth of phenomena on the earth's surface. Geological events echo throughout the landscape and throughout time, reverberating off one another. Yet in contrast to the pending doom we may associate with a mid-plate quake, in many cases such a confluence of events will give rise to global effects that are not only beneficial, but essential.

12

THE SWEAT OF THE EARTH

Over the past few years, oceanographer Edward Baker and his colleagues at the National Oceanic and Atmospheric Administration's Seattle labs have been systematically observing the Juan de Fuca Ridge, a plate boundary off the coast of Washington. Most of their missions have been routine, part of an extensive program to understand the effects of ridges on the ocean environment. But on one such mission they came across a startling anomaly.

The researchers were towing beneath the ship an instrument called a Sea Bird, which measures the temperature and chemistry of the water over a mile down. Amid the cool waters of the northeastern Pacific, the Sea Bird encountered something peculiar: a football-shaped mass of warm water some twelve miles across and nearly half a mile from top to bottom. At most, the mass of water was only a half a degree Fahrenheit warmer than the surrounding seawater. But the overall heat in this balloon—

which encompassed some twenty-four cubic miles of water—amounted to 18 billion kilowatt-hours of energy, or roughly enough to power a million households for a year.

The researchers dubbed this feature a megaplume. Judging from the minerals present in the mass of water, it appeared to have spurted out of the ocean crust. In addition, the researchers knew the megaplume couldn't have been there for long. In samples of the water taken from its midst, Baker and his colleagues found large sulfur-rich crystals that were expelled with the water. If these crystals had existed for more than a few days, they would have fallen to the ocean bottom or at least started to dissolve. Yet, as Baker says, "they looked like someone had drawn them on a piece of paper, just absolutely perfect, no pits, no dissolution, just perfect crystals like you would grow in a laboratory."

The ocean floor had let loose a sudden, massive burp of warm water, and almost as suddenly, it was gone. Two months later, when the researchers went back to look for the megaplume, it had disappeared without a trace. But a year later they found a second, smaller megaplume twenty-five miles north of where they had spotted the first. And two years after that, researchers in Japan reported a third megaplume above a ridge in the North Fiji Basin, convincing Baker and his colleagues that they had stumbled onto more than an anomaly.

"The oceanographic community hasn't looked at the worldwide ridge system in very many places in the detail that we have looked at it off the coast here," he says. "We haven't looked at it very long, and we found one and probably two of these things in the space of two years. And somebody in the western Pacific thinks they may have found one. So that leads me to think that these sorts of events are not particularly uncommon. There are sixty thousand kilometers of ridge around the world, and I'm sure that if only ten or five percent of it is undergoing active spreading from time to time, . . . a few of these things are popping off at least yearly, if not more than that, around the globe."

Such gross, episodic eruptions are still new enough that researchers don't completely understand their effects. As a result, geophysicists are still struggling to fit megaplumes into their picture of the earth. But Baker and his colleagues have proposed that these emissions of warm water may mirror thrusts of magma into the rock beneath them. A sudden injection of magma, they say, would cause the rock around the ridge to crack, releasing a torrent of heated water that lay trapped in the pores of the rock. If that's true, each megaplume may be a sort of smoke signal, alerting researchers to each surge of magma beneath a mid-ocean ridge.

This theory meshes well with researchers' current blueprint for the earth. In the eyes of geophysicists, hydrothermal circulation is inextricably intertwined with the pulsing bank of heat inside the planet. The megaplumes are just one more aspect of a continuum in the earth machine: from hot-spot plumes and churning mantle bringing heat to the surface, to the cooling effects of plate motions, to the waters of the world's oceans wafting that heat away. In fact, the last step is one of the most important.

Much as perspiration cools our own bodies, the sweat of the earth is whisking away the excess heat from its interior. The ocean deeps perfuse miles and miles of the mid-ocean ridge's cracked, young crust with their frigid water, forcing the water into the rock and down to the molten magma beneath the ridge, where it quickly heats up and races back to the surface, emerging at hydrothermal vents. Because cool water is continuously flushing through the rock, heat is dispersed much more efficiently. So efficiently, in fact, that of some 9 trillion calories of heat that radiates out of the earth each second, about a third is lost as a result of hydrothermal circulation instead of simple conduction.

This is the earth's artful way of shedding its heat. We already know that this discrepancy between oceans and continents may be the cause of the Supercontinent Cycle. Global "sweating" also turns out to have important consequences for life on the surface of the earth, a realization that has come to researchers in many different ways in recent years. But to understand these consequences, it helps to know the whole story of how researchers put together this particular image of heat dissipation from the first discovery of hydrothermal vents and the bizarre wildlife that takes advantage of these warm fountains.

Geophysicists had anticipated finding unusual features along the mid-ocean ridges since shortly after these plate borders were discovered. The hot, volcanic rock of the mid-ocean ridge is everywhere surrounded by water, and where these two elements exist side by side on the continents, the results can be spectacular. Witness Old Faithful and the more erratic geysers of Yellowstone National Park. Think also of the cauldrons of caustic fluids and the brightly colored, sulfur-rich rocks that give that park its name. Perhaps, geophysicists thought, the same extreme conditions existed at the bottom of the ocean.

In 1977, a team of researchers journeyed to the cobalt-blue waters of the eastern Pacific to look for such submarine hot springs. Their test site lay a mile and a half beneath them along a short, extremely active section of the world-girdling ridge system known as the Galapagos Ridge. A few years earlier, another group of researchers had discovered sudden jumps in the temperature readings of the waters just above this ridge. The blips appeared on three separate occasions, and each time the lowest instrument showed the biggest jump. Their results immediately raised hopes among geophysicists that this was at last what they had predicted they would someday find: warm jets of water squirting out of the ocean floor.

John Corliss, who was an oceanographer at Oregon State University, coordinated the mission back to the Galapagos Ridge. Then as now, Corliss was a merry, energetic fellow, but as head of the research cruise, he was also, by his own admission, a bundle of nerves. He and his colleagues were setting out to look for something that had never been seen. More than that, they were searching for a tiny trickle in an immense sea.

With that in mind, Corliss says, "I piled things on the ship." Every instrument that could possibly be of use to the mission found its way on board. One of the key instruments was a stroboscopic underwater camera known as ANGUS, for Acoustically Navigated Geological Undersea Surveyor. This rugged camera in a steel "gorilla cage" was developed for a project to map a piece of the Mid-Atlantic Ridge. Via acoustic signals, instead of heavy electric cable, it could continuously broadcast its position to scientists on board the ship and thus let them know exactly where they were photographing.

But the real workhorse of the Galapagos operation was the tiny sub called the *Alvin*. The *Alvin* is the lunar module of deep-sea exploration. Built around a titanium pressure-resistant sphere just seven feet in diameter, it can withstand the intense pressures at depths of over thirteen thousand feet below sea level. Since 1964, the *Alvin* has allowed oceanographers to do "field work" in the mysterious, buried terrain of the ocean floor. Those who have made the trip say that nothing compares to the intimate experience of creeping along the bottom in a submersible. When the little sub rendezvoused with their ship, Corliss and Tjeerd van Andel, a marine geologist now at Cambridge University in England, immediately went down to take a look.

The sub sank passively through the water column for about

an hour and a half, and on the way down visions of a strange otherworld streamed through the portholes. "The most spectacular thing was watching it get dark," Corliss remembers, "you know, deeper and deeper and deeper until finally it was just black. Then if we turned on the lights . . . you'd see things drifting in the water. We saw some interesting examples of a kind of snow—organisms that looked like big flakes of snow floating in the water. And if we turned the lights off, we'd see bioluminescent organisms flashing on. The sort of wave we made as we went down would disturb them and they'd flash."

Finally the *Alvin* reached the bottom, and the pilot jettisoned some of its weights, allowing the sub to hover above the ocean floor. The running lights were turned on, and the water's green glow poured in through the Plexiglas portholes. In contrast to the life the researchers had witnessed during their descent, the view was now a barren moonscape. Few creatures inhabit the ocean floor, where there is no light to sustain them.

The *Alvin* began inching across the ocean bottom. The sub's maximum speed is only two or three miles an hour, the same as a slow walk, and its circle of light extends only fifty feet. For all the fanfare about access to the deeps, exploring in the *Alvin* is nonetheless a fairly primitive affair. One can imagine this dive as the undersea equivalent of landing at night in the Grand Canyon and walking around with only a strong flashlight for guidance. The researchers were steering the tiny sub along sheer, high cliffs, the ridges that fence in the central ridge valley. Somber anticipation gripped them as they waited for a sign.

As always, the water at the bottom of the Pacific Ocean was frigid. "That day, I remember it was 2.038 degrees Celsius [35 degrees Fahrenheit]," says Corliss. "We were carrying thermistors, and I had a little alarm set up in the submarine. So I set the alarm to go off at 2.045—less than a hundredth of a degree higher. And when we first came upon the hot spring, the first thing I heard was the beeping sound. I started looking at the temperature device, and the numbers were going up—not spectacularly, a few hundredths of a degree. But then the pilot, Jack Donnelly, said, 'Oh, look, there are clams out here!' "

Stretched out before them was an oasis. Amid the barren ocean floor swarmed legions of bizarre, new animals. The creatures that the pilot first saw weren't actually clams, Corliss says; they were mussels. But as the researchers moved closer, they did see clams, luminous white creatures the size of dinner plates. The giant bivalves jammed the cracks between the black tufts of lava that covered the ocean floor. These rocks also stirred with novel species of eyeless shrimp, white crabs, translucent sea anemones and large, pink fish. Perhaps the oddest beasts in this oasis, though, were the tube worms. Their crusty white stalks, some several feet long, sprouted in dense bunches from the ocean floor; advancing and retreating from these hollow stalks were the blood-red worms themselves.

All of these strange, oversize animals clustered within a cloud of water that shimmered like haze above asphalt on a hot summer day. Clear, warm water was shooting up from every crack in a hundred-year-wide patch of rock. The *Alvin*'s thermistors showed the water peaking at sixty-three degrees Fahrenheit in some places.

Quickly the researchers discovered that ANGUS's camera could pick out the gleaming white clamshells with ease against the backdrop of the ocean floor. In the remaining days of the expedition, the researchers exploited this technique. They discovered four active hot springs: Clambake, Garden of Eden, Oyster Bed and Dandelions, each named for the distinct population of animals that thrived within them.

Once the researchers returned to shore, news of their findings spread rapidly through the scientific community. On the Galapagos Ridge, namesake of the nearby islands whose diverse wildlife inspired Charles Darwin's theory of evolution, oceanographers had discovered new islands of life that they had only begun to explore. And though the researchers had suspected all along that these hot springs existed, the real thing had far surpassed their imaginings.

Yet just two years later, in 1979, submarine explorers discovered an even more fantastic spectacle. Cruising in the *Alvin* along the crest of the East Pacific Rise just south of the mouth of the Gulf of California, the members of a joint American, French and Mexican research team came upon a hellish scene. Tall mineral chimneys like underworld organ pipes soared thirty feet above the ocean floor. Out of these pipes belched a cloud of hot, inky water, which billowed about the sub like the oily exhaust from a tailpipe. Black smokers, they were called, for not far away lay a beatific version of these vents. Milky waters gently coursed from a second set of worm-encrusted mounds, covering the ocean floor in a light, white sprinkling of dust.

Both the white and black smokers supported the same odd fauna that thrived at the Galapagos Ridge. But in contrast to those warm-water vents, these were a hundred times hotter. Thermistors showed the water emerging from the black smokers to be as high as 660 degrees Fahrenheit—hot enough to boil, if not prevented from doing so by the extreme pressures at ocean depths.

These black smokers, shooting out scalding water at speeds of up to sixteen feet per second, were truly the fire hoses of the deep. Or so researchers thought until Baker's megaplume discovery seven years later. It would take two thousand black smokers to expel the equivalent of a megaplume in a single year. But researchers believe the processes that create both of these warm jets are similar. Both sweat away the heat of the earth. And by ridding itself of heat in this way, the earth carries out a critical process: It maintains the fragile chemical balance that keeps us alive.

That's because, as water flushes through the rock at hydrothermal vents, it is not only heated; it is transformed. The cold, bottom water that percolates down into the cracks in the ocean crust carries its own complement of chemicals. But once that water is heated, many of those minerals precipitate out and are left behind in the ocean crust. At the same time, other minerals, such as iron and manganese, are picked up by the thermal water as it flows through the crust and are dragged up to the surface. The ocean water emerging from the vent thus has an entirely different chemical makeup than the water that went into it.

This simple set of chemical reactions resounds throughout the surface world. On a superficial level, it has given the undersea hot springs their otherworldly look. Because the minerals emerging from the vents dissolve only in superheated water, they quickly precipitate. "As soon as it hits the cold water, ffft, it all comes out," says geochemist Geoffrey Thompson of the fine-grained particles of iron sulfide that blacken the smoke at the ridge's hottest vents. Other compounds in the ejected water build up the tall, black smoker chimneys and white smoker mounds.

But the most abundant substance to emerge in the vent waters is hydrogen sulfide, the same stinky gas that makes the air around Hawaii's Kilauea Volcano difficult to breathe. All of the vent fluids are rich in this compound, which has the distinct odor of rotten eggs. When researchers on Corliss's cruise to the Galapagos opened up bottles of water from the hot springs, they were overcome by this unmistakable scent. Later, Corliss described it as the smell of success, for the researchers soon realized that this compound was supporting the diverse communities that populated the vents.

The giant clams, mussels, tube worms and other bizarre vent creatures subsist on a species of bacteria that metabolizes hydrogen sulfide. It ingests the compound from the vent water, snaps its chemical bonds and survives on the energy released. The vent animals, in turn, feed on the bacteria. Even tube worms, whose adult forms have no mouth or anus, survive in this way. While they're still young they take a few of the bacteria into their guts. Then the microbes continue to multiply in their stomachs long after the orifices have closed over.

These sulfur-metabolizing bacteria have captivated scientists, though not because they are particularly rare on earth. Researchers have seen their kind before in sewers and other places where organic matter is highly concentrated. But the prevalence of these microbes at undersea hot springs is intriguing because here the bacteria are the sole anchor for the food chain. In all other ecosystems the base of the food chain is made of creatures that carry out photosynthesis, turning the sun's rays into energy that other species can use. But the source of energy for this ecosystem is heat from the earth. The communities thriving at the vents are an anomaly of nature, a unique, self-contained ecosystem driven by the earth's internal fires.

It is therefore ironic that similar sulfur-loving bacteria, and not photosynthetic ones, are candidates for the original forms of life on earth. This is a theory that Corliss has championed almost since his original dives near the Galapagos Islands. "These hot springs were there from day one," he says. "From the time the ocean first cooled and rain fell on the hot crust and formed the oceans, the hot springs were there." Some 4 billion years ago, when the oceans were first forming, the hot springs may have been even more common than they are today, since volcanism was more intense in the earth's early history. Corliss sees evidence that the hot springs existed in rocks that are some of the oldest ever found: a 3.8-billion-year-old formation in southwest Greenland, known as the Isua series. These rocks, he says, contain the same chemicals that are supersaturated in the hot springs today.

But proving that life originated at hydrothermal vents involves more than just showing that it could have. Perhaps the strongest evidence for the theory comes from ongoing studies of bacteria. These simple, single-celled creatures are the most primitive forms of life on earth. Evolutionary biologist Carl Woese of the University of Illinois has created a tree of life that extends far enough down the evolutionary scale to include the bacteria. He has traced this lineage by comparing a specific RNA (ribonucleic acid) sequence within many different organisms. That sequence lies within a structure known as a ribosome, part of the cellular machinery for translating DNA into essential proteins.

According to Woese's evolutionary tree, the most primary of the present-day bacteria lie within a group called the archaebacteria. All archaebacteria thrive in intense heat, and most derive their energy from breaking chemical bonds. The bacteria that inhabit the undersea hot springs fall into this group. Woese believes that these modern species evolved from a similar type of ancient bacteria that probably grew at temperatures near the boiling point of water and may have drawn its energy from the sulfurous compounds spewing out of hydrothermal vents. If that's true, the bacteria seen at vents today could be the closest descendants we know of the original forms of life on earth. That would imply that the earth's unique way of venting its heat had a tremendous impact on that dawning era when life was first gaining its toehold on the planet.

Even if sulfur-loving bacteria were the first creatures to appear on earth, though, this is still not the complete solution to the origin of life. "There's still left the question of how does all that information get into the first little cell," says Corliss, who is now at the Goddard Space Flight Center in Maryland working on one aspect of this problem. How did life take the complicated step from a batch of chemicals in the prebiotic soup to an organized entity that could reproduce itself? That has proved to be the most difficult question for researchers to answer.

One contingent now believes that the earliest steps in this organization didn't take place either at deep-sea vents or in some "warm little pond," as Darwin imagined it, or even in the blistering-hot bombardment that is the modern-day image of our primordial earth. Having observed organic compounds in meteorites, these researchers believe the seeds of life may have been carried to earth preformed. In this scenario, the earth simply provided fertile ground for extraterrestrial imports.

That idea is made all the more intriguing by recent discoveries of bacteria thriving up to 1.74 miles beneath the earth's surface in sediments that are over 200 million years old. These deep microbes are a surprise and a thrill for scientists who had thought that life extended no more than about fifty yards into the ground. And they raise the question of whether such organisms were present in the ancient surface soil and gradually adapted as sediments piled on top of them, or whether they existed for billions of years in the deepest rock and migrated to the surface when conditions became hospitable, providing the foundation for life on earth.

So far no single theory on the origin of life has emerged victorious, mainly because there is little hard evidence to prove or refute any of them. For now the debate remains wide open. But whether or not hot springs eventually prove to be the cradles of all life, the reactions that take place there are ultimately important to the broader scope of modern life—not only the animals at deep-sea vents, but all the creatures that now inhabit the planet.

Salt of the Earth

"Probably what they're most important for, most people don't realize," says Geoffrey Thompson, standing in his Woods Hole lab amid row upon row of cardboard boxes full of rocks that he has scavenged from the deep-sea floor. As a geochemist, Thompson is attuned to the role that hot springs have played in the chemical balancing act that maintains our ecosphere. "They have provided us," he says, "with the answer to the question, Why does the sea remain salty?"

This age-old query dates at least as far back as the Icelandic sagas, written in the thirteenth century. Many people have heard some version of the story told there. A Danish king named Frodi is said to have possessed a mill that could grind out anything. The king ordered his slaves to grind out gold, peace and prosperity. But he would not let them rest, and when Frodi fell asleep the slaves ground out an army against him. The king was overthrown and the mill was pirated off to sea. Whereupon the thief ordered the slaves to grind salt, again without allowing them to rest. They ground so much salt that the ship, the thief and the mill all sank to the bottom of the sea. And that, the story goes, is why the sea is salty.

In geophysical circles, the question is not so much why the sea *is* salty, but why it *remains* just as salty as it is, no more, no less. For roughly three centuries, researchers have understood that the ocean's salts come from the minute amounts carried to it by rivers. These products of erosion include sodium, which combines with chloride ions in seawater to form common table salt, as well as magnesium and potassium.

Yet these minerals are not simply accumulating because, as researchers have known since the last century, the oceans have not been getting progressively saltier. Sedimentary rocks that formed billions of years ago are no less salty than those forming today. Somehow, salts and other minerals transported to the oceans must be removed at a comparable rate. But through the years geochemists had not been able to balance the input from rivers with the output from sedimentation and other processes at the ocean floor.

The discovery of hydrothermal vents and then megaplumes provided the missing details. Researchers now know that the ocean crust serves as a kind of global water-treatment plant. Salts are removed, pH is balanced, and various minerals are added as the water continuously flushes through the crust and out the vent stacks. According to one estimate, the entire ocean passes through this filter once every ten million years. Largely because of this multiplex reaction between seawater and ocean crust, the earth has maintained its chemical equilibrium. The oceans haven't turned into alkaline baths like Utah's Great Salt Lake.

Finally, this peculiar way the earth has of venting its heat has also influenced humanity in much more immediate ways. Through the burps of megaplumes and the steady puffing of hot, chemically altered water through the ocean crust, the earth concentrates its stores of precious metals. Select regions along the ridge axis become repositories for copper, silver, gold and zinc. Tectonic motions then conspire to move these remnants of the earth's internal activity onto land, where their bounty becomes available to us.

Such a store exists within the Troodos Massif, a six-thousand-foot-high phalanx of mountains on the island of Cyprus. Some 100 million years ago, a spreading ridge within the Mediterranean Sea was laying down new ocean crust. Later, the rotating African Plate pinched off the eastern opening of the sea, and the crust of the ridge was driven skyward to form the island range.

This heap of rock eventually became one of mankind's earliest sources of copper, the metal that took its name from Cyprus. Copper's burnished red color was readily visible in its raw state in the Troodos Massif, and this explains how preclassical people came to experiment with it. They discovered that this new substance could be hammered into tools with hard, sharp edges. These tools were far superior to the stone ones they had been using. By circa 2500 B.C., copper had come into general use throughout the Near East.

Over the history of mankind's development, this transition to metal weapons and tools proved to be a critical step. Like any act of history, it could have turned out differently. The Troodos Massif was not the only source of copper in the area, but it was one of the largest. And so, this sample of undersea hot springs shoved onto land takes its place in the story of our progress. Out of the ocean rose precious metals, out of the Stone Age rose civilization.

From a single act, a multitude of effects flow. Water courses through the ocean crust, and its impacts touch every corner of our existence. They speak to us of our origins, our survival and our way of life. Hydrothermal circulation is only one example. The dynamic earth has been spewing out the constituents of evolution and progress throughout its history, nurturing all species from the very first forms of life to our own. Thus every act on earth has a similar story to tell of both causes and consequences for the land, the sea and, as researchers have lately shown, the world's skies.

13

THE SHADOW OF A GIANT

Shortly after geophysicists evacuated Mount Pinatubo in the Philippines, the eruption began to make its presence felt around the world. Specks of dust, motes of ash and billowing clouds of bituminous gas had soared eighteen miles into the sky during the biggest blast. There, in the stratosphere, the particles were beyond the reach of rain clouds, which could have dumped them right back onto the ground. Instead, the ejecta set sail upon the high-speed stratospheric winds, and in the days and weeks that followed, researchers watched as the blast of gas and dust wrapped itself around the world's sky.

Just three days after the eruption, satellite images showed the cloud hovering ominously over India's Bay of Bengal. In two weeks the cloud had stretched westward across the Atlantic Ocean. In three, a layer of fine particles completely circled the equator. Come October, it had covered the entire planet and there it remained aloft for years.

Around the world, people could see the hazy skies that Pinatubo's remnant cloud produced, and particularly at the equator, where the dust was thickest, they were treated to months of spectacular sunsets. Yet the visible effects of this eruption were only the beginning. The tremendous outpouring also made a distinct impression on the earth's atmosphere, one that has finally given geophysicists the answer to a tenacious question about the way the world works.

Scientists have long suspected that volcanoes can affect the global climate. The first to make the connection between a major eruption and the weather was the statesman and philosopher Benjamin Franklin. Franklin's efforts to negotiate a peace treaty to end the Revolutionary War took him to Europe during the year 1783. And having been there on several occasions, he was among many to notice the peculiar blue haze or "dry fog" that cloaked the land that summer and fall. The following winter turned out to be unusually harsh. Soon thereafter, Franklin published an article that attributed these events to the eruption of Iceland's Laki Fissure, a great tear in the crust which had spewed out the greatest flood of lava in recorded history.

Since then, the theory has been suggested repeatedly as large eruptions have occurred. On April 10–11, 1815, the Tambora volcano erupted on the island of Sumbawa in Indonesia, expelling eleven cubic miles of magma in explosions of ash and fiery lava bombs. It cast such a shadow over the planet that New Englanders experienced snow in July, and in both the United States and Europe, the following year was known as "the year without a summer."

On August 26–27, 1883, Krakatau, an uninhabited island between Sumatra and Java, produced an explosive eruption that once again revived the theory. From a volcanic peak nearly 1,500 feet high, the island turned into a watery well 900 feet deep. The eruption was followed by one of the coolest years on record.

While scientists from Franklin on down through the twentieth century never entirely proved the connection between volcanoes and climate, most believed the dust and ash ejected from these volcanoes was cooling the climate by blocking the sun's rays. They thought these particles persisted in clouds in the stratosphere, a layer of the atmosphere that extends from about seven to thirty miles high. But then in May 1980, Mount St. Helens in Washington State erupted, and the ash-and-dust theory took a nosedive.

You couldn't have asked for a dustier volcano. The violent eruption that took place in the midst of an enormous avalanche blew off the top 1,300 feet of the mountain. Blistering-hot clouds of steam and pumice charged northward from the opened volcano at speeds of up to 250 miles per hour. The flows leveled 230 square miles of forestland and sent an estimated 500 million tons of ash as high as the stratosphere. It was the world's largest volcanic eruption in sixty-eight years, and the ash cloud circled the globe. But researchers found that it had no significant impact on the world climate.

Then two years later, Mexico's El Chichón volcano, which was previously thought to be extinct, produced an eruption that killed over 2,000 people. The volume of rock ejected was smaller than at Mount St. Helens, but its cloud of gas and dust reached 12 to 15 miles. And researchers expected its effect on the climate to be much greater.

By this time researchers knew that the clouds in the stratosphere that persisted after volcanic eruptions were actually made up of particles of sulfuric acid, not ash or dust. Researchers believed that sulfur dioxide emitted by volcanoes would turn into sulfuric acid in the stratosphere, creating a haze of acid droplets that would block the sun's rays and keep them from warming the air below. El Chichón's ejecta was particularly rich in sulfur dioxide, whereas Mount St. Helens's was not. All told, El Chichón spewed out about 7 million tons of sulfur dioxide gas, and thus geophysicists sat in wait for a bout of global cooling.

Computer models predicted that El Chichón would reduce the earth's temperature by roughly half a degree Fahrenheit. There is no doubt that it had some effect. After the eruption the opaqueness of the atmosphere increased around the globe. But the measurement of global cooling was complicated by another weather event which also took place that year: El Niño.

El Niño is an atmospheric phenomenon that springs up in the southern Pacific every three to seven years. The normal trade winds along the equator temporarily lapse or even reverse direction and wreak havoc upon the weather and ocean currents. These, in turn, upset the local marine life as well as the fishermen who depend upon them, particularly in the case of the El Niño that struck in 1982–83. That was one of the most severe in the last hundred years. Its effects showed up all around the world, from droughts in Australia to heavy rains in California. And like most El Niños, this one also had a temporary warming effect on the global climate. As a result, geophysicists couldn't say exactly how much El Chichón had cooled the atmosphere.

Not until Pinatubo erupted did researchers have a chance to see once and for all whether volcanoes affect climate. Pinatubo shot almost three times as much sulfur dioxide into the stratosphere as El Chichón, about 20 million tons in all. Shortly after the eruption, James Hansen, at NASA's Institute for Space Studies in New York, began to estimate the effects to come from these sulfurous emissions.

Hansen is well known as the man who has brought the

concept of global warming to the front pages. Since the early 1980s he has argued with increasing forcefulness that global warming from man-made emissions of carbon dioxide and other greenhouse gases is already showing up in the climate record. Until recently he was convinced that the warming in the 1990s would be so noticeable that those colleagues who had been reluctant to accept his global-warming scenario would be forced to recant. One of Hansen's fellow atmospheric researchers has suggested that Hansen may have been so quick to examine the impact of Pinatubo on the global climate because he didn't want his critics to use the lower temperatures caused by the volcano to accuse him of crying wolf about global warming.

But Hansen is also an old hand at exploring this question of how volcanoes affect the world's climate. Several years ago, he used a simpler version of his climate model to test whether it could reproduce the climatic effects of a 1963 eruption of Mount Agung in Bali, in which gas and dust reached the stratosphere. Hansen concluded that the predicted cooling was "consistent with observations," but the temperature change was not large enough to stand out over the normal variations within a given year. Pinatubo, on the other hand, was enormous. When it erupted, Hansen's group at Goddard quickly recognized that the ejecta would provide "an acid test" of their model's ability to predict climate changes.

Into the computer the researchers put the best estimates for the size, thickness and global distribution of the aerosol particles, and after crunching through the numbers, the climate model supplied a prediction. Pinatubo was expected to cool the climate in 1992 and 1993, reaching a minimum late in 1992 of about 0.9 degree Fahrenheit below the average global temperature prior to the eruption. "The predicted cooling is sufficiently large," Hansen and his colleagues wrote in January

1992, "that by mid 1992 it should even overwhelm global warming associated with an El Niño that appears to be developing."

From then on, it was just a matter of watching and waiting. Pinatubo was one of the most closely monitored eruptions of all time. Researchers observed it from the ground, from airplanes and from orbit. They measured the drop in solar radiation coming through the stratosphere as well as the rise in sunlight reflected back into space. Sure enough, by the middle of 1992, the weather satellites and weather balloons already showed the effects of that lost sunlight. The first half of 1992 was 0.7 to 0.9 degree Fahrenheit cooler worldwide than the first half of 1991. The decline in global surface temperatures continued throughout the year, eventually bottoming out around December 1992 at 1 degree Fahrenheit below the average of the year before Pinatubo's eruption.

The bulk of that cooling took place beneath the thick haze that straddled the equator: in the Amazon and Congo basins as well as the equatorial Atlantic and Pacific. Even though an El Niño did indeed spring up in the early months of 1992, it was a mild one, and thus its warming effects, which are usually strongest in the equatorial Pacific, only briefly reversed the cooling trend wrought by Pinatubo.

Hansen had put his climate model to the test, and it proved to be accurate, at least on this particular question. But of much broader interest is the fact that a long-held theory was confirmed. Volcanoes can cool the global climate. What's more, Hansen now says that the warming trend which he believes was in effect before Pinatubo erupted won't become generally obvious for several more years as a result of the volcano's effects. If that is true, then Pinatubo has given us all a slight reprieve.

Weakening of the Shield

Yet the emissions from Pinatubo may also have had adverse effects on the atmosphere. The 1992 and 1993 Antarctic ozone holes were deeper and around 20 percent bigger than in previous years. Susan Solomon at the National Oceanic and Atmospheric Administration's aeronomy laboratory in Boulder has argued that Pinatubo's aerosols probably contributed to the Antarctic phenomenon. The same sort of ozone-destroying reactions that take place on stratospheric clouds at the poles occurred on its sulfuric acid clouds.

More disturbing, because more people live under its shade, is the fact that the worldwide ozone shield also reached a record low in the second half of 1992 and early 1993. Ozone has been decreasing gradually all over the globe by about 3 percent a decade. But James Gleason and his colleagues at NASA's Goddard Space Flight Center and elsewhere, who man the Total Ozone Mapping Spectrometer, or TOMS, satellite instrument, saw almost that much of a drop in a single year. Though many ozone researchers were expecting a massive impact on ozone as a result of Pinatubo, they didn't expect a decline quite so severe.

Outside the polar regions with their ozone holes, Gleason and his colleagues found the average level of ozone across the globe to be 2 to 3 percent lower in 1992 and 1993 than in any of the preceding thirteen years. The drop began in March 1992 and continued to a low of 4.7 percent below average throughout the first half of 1993. But when the researchers looked at where the ozone levels were dropping, the results were even more confusing.

The ozone level actually rose in the tropics. But the drop at middle and high latitudes compensated for it. That drop was particularly large in the Northern Hemisphere, where the bulk of the world's population resides. Over Europe and North America, ozone was 9 percent below normal in the spring of 1993, which corresponds to roughly a 12 percent increase in harmful ultraviolet rays.

Everyone knows about the harmful effects of increased UV. Everything from the eyes and soft noses of seals to the flesh of plants to our own exposed skin is in the line of fire. They are all vulnerable to cancer and other harmful genetic mutations induced by the UV rays.

As little as researchers know about why the ozone is thinning globally, this precipitous drop after Mount Pinatubo's eruption only adds to the mystery. Some of the first-generation models that researchers have cobbled together reproduce the uneven quality to the ozone losses: the increases in the tropics and decreases in the middle to high latitudes. But these models are only works in progress, since the highs and lows are still much too large.

While geophysicists may not be sure of all the ways that Pinatubo affected the atmosphere, they are sure of the factors that led to most of those impacts. This eruption was unique in having unleashed its contents into the skies so explosively and with so much sulfur that it could cause a temperature change. And both of those traits originated where the magma was being held, deep inside the volcano.

By examining the rocks that flew out of Pinatubo and reconstructing the blast, researchers hoped to connect the impact this volcano had on the sky with its origins in the deep earth. In the process they gained some insight into the unusual mix of activities that leads to such a violent outburst. They discovered a strange new picture of the magma pool beneath volcanoes like Pinatubo and uncovered a recurring trigger for the eruption. Yet their newfound insight has only served to deepen their sense of the mysteries that still remain.

What Made This Volcano Such a Corker

Since it was the sulfur that caused the global atmospheric cooling, and probably the ozone losses, researchers made a special point of tracing it back to the rock in which it was dissolved prior to the eruption. "The problem is that when we do that for sulfur in the case of Pinatubo," says volcanologist Terrence Gerlach at the U.S. Geological Survey in Vancouver, "it doesn't even remotely match the amount of sulfur that actually came out of that eruption. We can't account for it."

Nine years before the Pinatubo eruption, researchers discovered exactly the same discrepancy at El Chichón. Since then this nagging dilemma has come to be known as the excess sulfur problem, and geophysicists have found their ignorance growing more and more embarrassing as the years have progressed. Just as atmospheric scientists were beginning to grasp the importance of a volcano's sulfur dioxide emissions, solid-earth scientists were finding that they had no idea where all this sulfur was coming from.

They knew that the hydrothermal circulation at ridges was loading the ocean crust with volatile substances—primarily water, but also sulfur, chlorine and smaller amounts of carbon dioxide. As the ocean crust goes down into the trench, the volatiles go with it. Then amid the heat and pressure around sixty miles down, geophysicists believe the volatiles escape from the slab and dissolve in molten rock, which rises back toward the surface. When the volcano cracks open, it's like popping the cork on a champagne bottle. The volatile gases come out of solution and begin to boil explosively, causing the rock around them to shoot out of the ground like warm bubbly.

The Pinatubo eruption was so explosive that researchers who have studied the Philippine outpouring say there were no lava flows to speak of. Most of the magma came out in the form of flying chunks of pumice. Embedded within these burning chunks were microscopic bits of rock that still contained the sulfur and other volatiles which were dissolved in them when they were molten magma. But the amount of sulfur that researchers found in these so-called melt inclusions was a far cry from the amount that was blasted into the sky.

Recently, Gerlach and Henry Westrich at Sandia National Laboratories in Albuquerque, New Mexico, have put forth an interesting solution. After probing Pinatubo's rocks, the two researchers confirmed a hunch they had formed from looking at other eruptions. That hunch gives us an unusual image of what lies beneath volcanoes. In short, Westrich and Gerlach think the sulfur at Pinatubo was trapped as a gas underground. In other words, it was not dissolved in the molten rock but physically displaced the rock. That would resemble large pockets of gas floating around a sealed bottle of champagne. The subterranean gas bubbles would presumably exist even at great depths where the pressures are quite high—a bizarre image that scientists were reluctant to accept at first.

"Part of the problem is that most of the people studying these things are studying them like geologists usually do," Gerlach says. "Basically something is either mineral matter or chilled liquid. You can't, however, see the gas phase which may have been present because it's gone. And it's only in recent years, when people have begun to use more of the remote-sensing technology, where we actually measure the gaseous emission rates from an erupting volcano, that we have been able to get this kind of data."

Yet even gas bubbles can lie in wait underground for many years, and the geological evidence at Pinatubo indicates that this was the case. Pinatubo was slowly building toward an explosive eruption. But as John Pallister at the U.S. Geological Survey in Denver has shown, this volcano wouldn't have blown in 1991 had it not been for another factor.

About a year before the eruption, the volcano became a little more active. Some of the pores on the north side of the volcano started jetting gas. Then on April 2, still several months before the eruption, came the first of many steam explosions. Looking back on that increased activity, researchers now believe it was a symptom of what was happening inside the volcano. That's because the magma that eventually shot out of Pinatubo was composed of two different types of rock, a whitish pumice with clods of black basalt embedded in it. In order for these two rocks to be mixed, magma deep inside the earth had to mix. Researchers believe that basaltic rock from twenty to thirty miles deep in the earth, possibly from the melted ocean plate itself, rose to join with the shallow pool of sulfur-rich magma.

"So it was sitting there, in effect supercharged and ready to go, and here comes this hot basalt along," says Pallister. As he sees it, the basalt added heat and probably some more volatiles to the mix. It stirred the magma, starting a chain reaction in which the champagne bubbles of gas started to boil, and the volcano burst wide open. But it was this intrusion from below that triggered the eruption.

Even more eye-opening, from his fieldwork at Pinatubo, Pallister has discovered that the volcano has erupted in much this same way every 400 or 500 years for the last 30,000 years. He and other researchers envision the volatile gases gradually building between each blast. But time and again it has taken this kind of intrusion from the deep to finally make Pinatubo blow.

"Does that mean that Pinatubo has had such a profound impact on the atmosphere multiple times in the past?" Pallister asks. "Has it erupted sulfur-rich magmas all these times?" And what about other volcanoes? "Why would El Chichón and Pinatubo be so sulfur-rich and other eruptions from subduction-related volcanoes, a good example being Mount St. Helens, be very sulfur-poor?" asks James Luhr at the Smithsonian Institution in Washington, D.C., who first encountered this question in his fieldwork at El Chichón. "That's one of the million-dollar questions that everyone is worrying about."

Since volcanologists don't know what makes a particular volcano erupt massive amounts of sulfur, they can't say whether these eruptions were anomalies or relatively common phenomena that have occurred repeatedly through geologic history. They can't say how often our skies experience this barrage from the interior. There is the fact, however, that two high-sulfur explosive eruptions have taken place within a span of just ten years. This suggests a fairly frequent cycle.

But even to test that idea is, so far, impossible. Researchers can't dissect rocks from prior eruptions to predict whether future eruptions are going to be sulfurous, because the hard evidence of that sulfur has disappeared. In what Luhr calls "a cruel trick of nature," the one known fingerprint of a sulfurous volcano is a compound called anhydrite, which dissolves in groundwater. Just a few years after an eruption—a mere geological instant—all traces of anhydrite are gone. This explains why the substance had pretty much eluded scientists before it was discovered at El Chichón.

Eruptions of Old

The inability to answer these questions has broader repercussions. The effects of Pinatubo, the largest volcanic eruption of the century, lasted several years. As of October 1994, the ozone layer was beginning to show signs of recovering from its record lows, and the cooling influence of Pinatubo's aerosols on the surface appeared to be waning. But how does Pinatubo's eruption compare to the last century's massive explosions?

In 1883, Krakatau launched a cloud of ejecta fifty miles into the air—about 25 million tons in all. Yet even this pales in comparison to the eruption of Tambora in 1815. This volcano produced five times the ash of Krakatau and put six times as much gas and dust into the stratosphere. The once 4,300-meter-high stratovolcano lost 1,400 meters of its summit cone. By these standards, Pinatubo was just a bush-leaguer.

In addition, such gigantic eruptions are known to have occurred throughout history. Ancient accounts of sustained periods of cold and dark exist in Egyptian papyruses as well as biblical, Sumerian and Mayan literature. Geophysicists have even found evidence for a major volcanic eruption in A.D. 536. Though they have no idea where it erupted, its climatic effects were described by a contemporary historian and geographer. From his accounts, researchers know that the volcanic haze from this giant eruption nearly eclipsed the sun's rays and wreaked havoc on the climate for many years.

Then there was Toba, on the Indonesian island of Sumatra, which erupted 73,400 years ago. It produced 670 cubic miles of material, compared to only 11 cubic miles for Tambora. About a billion tons of sulfurous gas were spewed into the stratosphere, making this one of the largest explosive eruptions ever.

Stephen Self at the University of Hawaii in Manoa and Michael Rampino at NASA's Goddard Institute for Space Studies in New York have proposed that Toba caused an average global drop in temperature of seven to nine degrees Fahrenheit, compared with just one degree for the Pinatubo eruption. They have even suggested that this volcanic cooling amplified a global ice age that was already under way.

On the large end of the scale, volcanic eruptions can approach such size that they may have an impact on the thin skin of life that rests upon the globe. For some years now, geophysicists have been kicking around the idea that such mega-eruptions cause plant and animal extinctions. Just as the earth created a hospitable platform for life, so may the earth have repeatedly taken life away.

Thus as geophysicists have attempted to connect their knowledge of the dynamic interior with the world at the surface, they have begun to see the impact such immediate, catastrophic events as volcanoes might have when repeated throughout history. Similarly, scars in the continental crust mine the ground that we walk on, and the steady cooling of the earth maintains the ocean realm. Though mid-plate quakes, megaplumes and the future effects of any one volcano still present mysteries for researchers to solve, the idiosyncrasies of the crust are beginning to take their place in the complex tapestry of the planet. Gradually it is becoming easier for geophysicists to hold the world in their arms.

Yet even as the varied threads of the earth come together, geophysicists are realizing that some of its activities go beyond the fabric of interactions they are weaving. As researchers have grappled with explanations for such events as mass extinctions, they have also come to understand that they cannot limit themselves to the forces of the churning interior. No picture of the working world is complete even once all the layers from the core to the atmosphere have been taken into account, for the earth does not exist in a vacuum.

14

LIFTING OFF

It is easier to believe that Yankee professors would lie, than that stones would fall from heaven.

—Thomas Jefferson, *reacting to the news that two Yale professors had reported the fall of a meteorite*

The earth revolves against the backdrop of space, exposed to sizable bodies of matter that occasionally hurtle into its neighborhood. From time to time one such body streaks through the earth's atmosphere and bombs the planet. Farmers have been unearthing meteorites in their fields and quarriers have plucked them from their quarries since as early as 2000 B.C. In spite of these discoveries, scientists and laymen alike have consistently scoffed at the notion of stones falling out of the sky. Well into the twentieth century, geologists proclaimed that meteorite impacts were so rare as to be insignificant. Any mention of their possible effects on the earth was met with outright hostility.

Only after researchers physically journeyed off the earth were they able mentally to get beyond its bounds. During the series of Apollo moon landings in the 1970s, the astronauts brought back samples of the lunar surface. From these, researchers determined just how often the moon suffered from asteroid impacts. The results convinced planetary scientist that the universe was indeed a very violent place. But even then many geophysicists didn't take the lesson to heart. The serious study of asteroid impacts on earth had to wait for another push.

That came in 1980, when a group of researchers proposed the theory that an asteroid impact killed off the dinosaurs 65 million years ago. In fact, the dinosaurs were only the most famous of the life-forms that became extinct at the end of the Cretaceous period. More than half of all species of plants and animals that existed at that time disappeared. Their fossil remains cease appearing in the rock record. And right at that point in the geologic strata all across the world lies a layer of clay less than half an inch thick.

In the late 1970s, the father-and-son team of Luis and Walter Alvarez and their colleagues at the University of California at Berkeley began to study this clay layer in the mountainside outside the medieval town of Gubbio, Italy. The researchers soon discovered that the clay contained an unusually high concentration of iridium. This metal is rare in the earth's crust but common in comets, asteroids and other extraterrestrial objects. After much debate over the source of this iridium, the researchers went public with a radical theory.

They proposed that an asteroid about six miles in diameter had slammed into the earth, kicking up a cloud of iridium-rich dust. That dust eventually settled into the clay layer that punctuates the tail end of the Cretaceous. But before it did, the researchers believed, it would have cast a pall across the globe. They proposed that the impact on climate and on photosynthesis, both in the ocean and on the land, could account for all the plant and animal extinctions that followed.

The Alvarez theory wasn't exactly warmly embraced. Besides the fact that the researchers could produce no crater as evidence of the impact, another competing theory did have physical evidence. The massive Deccan eruptions in India also took place around 65 million years ago. Many geophysicists believed this volcanic upheaval could have created a global environmental nightmare far greater than the one caused by Mount Pinatubo, resulting in the same cooling and disruption of photosynthesis that the Alvarez theory attributed to an asteroid impact. Throughout the 1980s the debate over the dinosaurs' extinction swung between these two theories of an internal versus an extraterrestrial source for the catastrophe.

Then in 1990, in an unusual series of events, the impact theorists found their crater. They had narrowed the search down to the Caribbean–Gulf of Mexico region through the exhaustive study of tektites—glassy spherules that the asteroid would have melted and kicked up on impact. At that stage a reporter for the *Houston Chronicle* mentioned to crater hunter Alan Hildebrand, then a Ph.D. student at the University of Arizona, that he had worked on a story nearly a decade earlier about a possible crater discovered in Mexico's Yucatán Peninsula.

The crater, known as Chicxulub for the town nearest its center, was in just the right place, and at 186 miles wide it was large enough to uphold the impact theory. In fact, Chicxulub is the largest impact crater that anyone has found on earth so far, and one of the greatest in the solar system. When researchers finally determined the crater's age in 1992, that also turned out to be just right: 65 million years old. The date of the impact convinced almost all remaining doubters that the Chicxulub asteroid was indeed the killing blow to the lost Cretaceous species.

What is still being debated is whether other factors, such as the Deccan volcanism or, more commonly, other impacts, played a role in the extinctions. Lately some researchers have been arguing that what hit the earth 65 million years ago was more like an asteroid shower. Multiple impacts may have occurred at roughly the same time to cause the global extinctions.

To prove that, researchers will have to find multiple craters that date to the end of the Cretaceous. But in recent years new impact structures have been showing up at the rate of two or three a year. In part, this wealth of new finds stems from the fact that researchers now have a much better idea of what to look for. Yet it is also a clear sign that geophysicists have recognized how much the planet's nature stems not from its own internal gyrations, but from the slings and arrows of the universe around it.

That realization has thrown researchers into much the same situation they were in when hot-spot plumes first became a hot topic. The success of the Alvarez theory has inspired a flurry of new impact scenarios. Some of the most interesting of those theories have asteroid impacts impinging on the deepest layers of the earth. They force geophysicists to take another look at the relationship between impacts and internal processes. Rather than competing to direct the planet's evolution, they may be cooperating.

In fact, we have already encountered one example of that in an earlier chapter: the idea that the moon was splashed out of the earth when a giant planetesimal collided with the planet early in its history. A by-product of that collision may have been global-scale melting, perhaps resulting in a much hotter core today.

Another interesting possibility has been raised by Michael Rampino and Richard Stothers at NASA's Goddard Institute for Space Studies in New York. They have shown that large impacts recur on earth at roughly 32-million-year intervals, and so do bouts of flood volcanism as well as mass extinctions. Rampino and Stothers have proposed that these events are all connected. The impact of a large object, they say, could cause a hot-spot plume head lying just beneath the earth's surface to suddenly decompress and let loose an enormous flood of magma. So a feature like the Deccan Traps might actually be covering the site where an asteroid once hit. Though many researchers question whether this proposed scenario could work, Rampino and Stothers have managed to link the two competing extinction theories under the rubric of "impact volcanism."

For the last few years, a handful of researchers have also been arguing another radical idea. They believe that impacts are the cause of magnetic-field reversals, and they have invented a variety of scenarios for how this would work. Some researchers have the impact simply jolting the molten iron in the core into a sudden change in circulation. Others have it triggering the liftoff of many plumes from the core-mantle boundary, which would, in turn, change the way the core mixes. Perhaps the most elaborate scheme has the impact causing cooling on the earth's surface and increasing the size of the ice caps. This would result in a slight change in the way the solid crust and mantle spin, causing a shift between them and the molten core and thereby disrupting the iron currents that generate the field.

It is an unfortunate setback for these theories that reversals and impacts don't match up as nicely as researchers first thought they did. A case in point: the impact that killed the dinosaurs has no reversal associated with it. Such exceptions don't necessarily rule out the theory, since not every impact would have to cause a flip. But so far this idea is not well substantiated.

Indeed, the spate of asteroid theories in recent years is beginning to give way to a backlash, much the way hot-spot plume theories once did. Researchers are forcing theorists to confront the accumulated evidence that contradicts their ideas. This tug-of-war is by no means the end of impact theories. It is merely the scientific process. Over the next few years, researchers will slowly separate the wheat from the chaff. And when the dust has settled, the Alvarez theory and perhaps some of the others will be incorporated into the picture of the earth.

In thinking about the planet as a whole, geophysicists are thus coming to grips with the importance of the solar system around it. Adopting this distant vantage point is in many ways an apt conclusion to their efforts over the last decade. Researchers are now comparing what is known of the earth with other planets to see how they have fared under similar conditions.

Only one other planet bears a real resemblance to the earth. Our sister planet Venus is similar in size, age, density and distance from the sun. Unlike the other terrestrial planets, it is massive enough that its interior has not yet cooled and solidified. Mars, Mercury and the moon are all thought to be geologically dead. But Venus is very much alive.

Over the last few years, geophysicists have watched as the satellite *Magellan* has sent back extraordinary pictures of the surface of Venus. *Magellan* used radar to penetrate the dense and extremely hot clouds of Venus's atmosphere. From its first encounter with Venus in August 1990 to its final plunge toward the planet's surface in October 1994, it provided researchers on earth with convincing evidence that the same churning that goes on inside our own planet is also at work inside Venus.

Magellan scientists have seen thick pancakes on the Venusian surface, where viscous lava has flowed like toothpaste out of the planet. Before the mission, Venus's crust was also thought to be much more ductile than the earth's as a result of the nearly 900-degree-Fahrenheit atmosphere. And that conclusion is borne out in the *Magellan* images. "Something from the inside has squeezed the surface around as if it were taffy," says Gordon Pettengill at MIT's Center for Space Research. Venus is scarred with enormous rifts where the crust has been pulled apart, and linear mountain belts where the hot, perhaps fluid rock in its interior is pinching the crust from underneath.

What really stands out about Venus's surface, though, is how smooth it is. The number of craters is relatively small, even taking into account that the dense atmosphere filters out most of the smaller asteroids. And the average age of those pockmarks is only 500 million years, a tenth of the time that Venus has withstood this asteroid rain. Thus geophysicists have come to believe that the activity inside Venus must have resurfaced the planet fairly recently.

On the earth, impact craters are continuously being erased. Rain erodes most of the craters on land, and the entire crust of the ocean is resurfaced roughly every 100 million years as it disappears down the ocean trenches. But Venus doesn't have rain to erase its pockmarks. Nor do most researchers believe that it has tectonic plates. The thin Venusian crust appears to be a single, all-encompassing layer of rock.

Nevertheless, Venus does have a hot, churning interior like the earth's. Several researchers have proposed that 500 million years ago Venus experienced an internal upheaval. Some catastrophic event, perhaps resembling the flushing event in the earth's mantle, sucked the crust down into the planet and left a fresh, unblemished surface in its wake. Other researchers argue that the repaving was the result of a more gradual process. As Sean Solomon, director of the Department of Terrestrial Magnetism at the Carnegie Institution of Washington, D.C., has recently suggested, the interior of the planet may simply have cooled to the point that it lost the ability to vigorously knead the surface crust 500 million years ago and thus stopped erasing the craters.

What is clear is that in spite of the many similarities between the two planets, Venus doesn't work in quite the same way as our own. For some reason it never developed the tectonic plates that have so much to do with how the earth looks. Its surface interacts differently with its interior. Thus even in comparison to this sister world, our home planet is exceptional.

Studying its particular quirks is the only way to predict what is in store for the earth. So far its evolution has been directed by both internal and external factors. They have brought the planet from its initial accretion through a fiery adolescence to its current lush state. And if the present is the key to the past, so, too, the past is the key to the future. Can our insight into the past tell us what will be happening on earth millions of years from now?

Paul Hoffman posed an answer to this question in the article "Supercontinents" in the *Encyclopedia of Earth Systems Science*: "It is evident from plate motions over the last 100 million years that a future supercontinent is gathering around Eurasia. . . . If the Pacific Ocean continues to close (as seems likely if South America follows North America in overriding the East Pacific spreading-ridge), then the American continents will finally join the future supercontinent about 100 million years from now."

With the dispersal of this supercontinent may come worldwide flooding. With that may come global warming. The next major greenhouse episode lies more than 100 million years in the future—that is, if the pressure that we are putting on our environment doesn't bring about these conditions much sooner.

Man's return of fossil carbon into the atmosphere might cause an artificially induced warming in the next century.

As instruments of global climatic change, we humans now rival such cataclysmic events as supercontinent breakups and asteroid impacts. There is no escaping the fact that our tenancy on this planet is not guaranteed. Recognizing that, both scientists and citizens have begun to take action. As geophysicists attempt to understand how the earth maintains its fragile environmental balance, a handful of agreements from the international community are forcing the people of the world to act upon those findings so that we may exist upon the earth for ages to come.

But geophysicists also know that it is of little consequence to the planet whether we flood or fry ourselves out of existence. Efforts to "save the earth" should properly be termed "save our species." With or without our survival, the earth will continue for eons. Fueled by heat from within, the plates will keep moving, the mantle will churn, and the continents will continue to trap heat beneath them. Heat from the core will keep the earth's magnetic field alive, and hot-spot plumes will ferry their magma skyward.

Gradually all of these processes are cooling the earth. The mixing of the mantle and with it the motions of the plates are ever so slowly winding down. As a result, the number of supercontinents over the next four and a half billion years will probably be fewer than in the past. Some researchers say the plates might even stop moving altogether in the next one or two billion years.

Whether or not the earth will still be the platform of humanity as these long-term changes take place is an open question. We are pushing out into space at a steady rate, and many people hold the hope that other worlds will be our future. That possibility gives us an unusual sense of our stay on earth as temporary. The home planet was simply the place where we grew up before leaving to explore the universe.

If we want to be so bold as to imagine that humans will survive billions of years into the future, then our journey will have to take us beyond the bounds of this planet. The stage is already set for the earth's theatrical finale. Our sun is slowly evolving into a red giant star. Five billion years from now, its fiery cloak will extend nearly to Mars. And the earth, literally inside the sun, will vaporize. In the end its remains, and with them our own, will spread throughout the galaxy whence they came, dust to dust.

INDEX

Index

Index

Index

Index

Index

Index

Index

Index